Quick Start Guide to Industry 4.0

One-stop reference guide for Industry 4.0

My eyes established a vision for my life,

After seeing the beauty of Lord Shri Krishna (with belief, devotion and love).

My senses started working intelligently,

After understanding the morals of Lord Shri Krishna's stories and his past times (with belief, devotion and love).

My hands started writing good books,

After folding hands in front of Lord Shri Krishna (with belief, devotion and love).

Oh Lord Krishna, what am I without your blessing?

Oh Lord Krishna, your blessing is the reason for my breath, for my actions, for health, for prosperity, for peace and everything that I possess.

Hare Krishna Hare Krishna, Krishna Krishna Hare Hare,

Hare Rama Hare Rama, Rama Rama Hare Hare,

***Jai Sriman Narayana** !*

Dedicated to Lord Shree Krishna!

Table of contents

Disclaimer

My best efforts have been made to validate the content provided in this book, the author does not assume any responsibility for errors, omissions, or contrary interpretations of the subject matter contained within.

The information provided in this book is for educational purposes only. Author does not accept any responsibilities for any liabilities or damages, real or perceived, resulting from the use of this information.

Preface

If we look back at the history of industrial revolutions, we will have to start with first revolution which used hydraulics and produced steam engines between 1700 and 1800's. Later was the second industrial revolution which used electricity and produced products in the 1900's. Then there was third industrial revolution which used electrical and electronical engineering with the help of computers and robots for production of machines and industrial operations.

Now the fourth industrial revolution is Industry 4.0, which is the combination and aggregation of numerous IT technologies, processes, and machinery for faster operations making it more agile and impeccable. Industry 4.0 or Fourth industrial revolution implements big-data, cyber-physical systems, cloud technologies, sensors, robotics, and data management technologies. With the help of these technologies it makes the machines, infrastructure and the complete industry smarter. Not only that, it also reduces the time, labor, and enables to make more effective decisions in industrial operations. Industry 4.0 will not only improve effectiveness, efficiency, quality and flexibility to factories, but will also improve the operations at suppliers' side by providing the necessary and accurate information at the right time to right stakeholders improving innovation.

Realizing that there are very few books on industry 4.0, here is my humble effort to give an overview on Industry 4.0, its components, principles, approach for setting up smart factory, and many more interesting topics.

I should also tell you that this book will be useful for beginners, mid and top-level management like managers, consultants, etc. to give a holistic understanding on industry 4.0.

Next, this book has not been written just for the sake of writing doing 'beating around the bush' on the same topics with some rhetoric paragraphs and images (and explanations) just to increase the number of pages. The only intention I had while writing this book was to give short, precise and pragmatic knowledge which I have researched and gained in my experience.

Acknowledgments

If I have to start thanking people name by name,

I would first thank my mother Krishna Kumari Pabbathi for giving birth to me with all my organs and body in good condition.

I would thank my father Surya Prakash Pabbathi for teaching me the right morals, values, and placing me in one of the best school while sacrificing his comforts and pleasures.

I would thank my grandfather Satyanarayana Pabbathi who iterated that good books are the greatest friends and exemplified his life for principles.

I would thank Balasaraswathi Penmetsa for teaching Mathematics patiently in my 10th grade class, when I was poor at Mathematics, and when no professional teacher could teach me with patience.

I would thank my graduation friend Venkatesham Padigela who inspired me to become a topper in my graduation.

I would thank my first spiritual alma mater - ISKCON and my first spiritual mentor Kalakanta Prabhu for making me understand life and Lord Krishna (which is best thing that ever happened in all my life).

I know that I am still missing numerous people who have taught, helped, and supported me numerous times in my life.

Hence, I would give my highest respects and would surrender to the Supreme Lord Krishna for arranging all these wonderful people in my life, and for giving me so many lessons, experiences, and knowledge!!!

About the author

Kiran Kumar Pabbathi has worked for various companies in the IT industry which gave him detailed insight of ITSM, ITAM, IAM, Cloud, IoT and EAM best practices.

Kiran has had the privilege to work in different roles taking care of service desk operations, request fulfillment, incident management, problem management, change management, configuration management, sharepoint administration, project management, ITIL consulting and trainings, and ITAM consulting.

Kiran is a certified professional in ITIL® Expert, PRINCE2® (Foundation and Practitioner), ISO/IEC 31000 and ISO/IEC 27005 Certified Risk Manager, Six Sigma Green Belt, ISO/IEC20K – Foundation, Cloud Computing – Foundation, TMAP – Foundation (Test Management Professional), MCP in SharePoint 2003 Customizations, and MCTS in MS Project 2007. Kiran is also an accredited ITIL trainer and Invigilator from BCS, UK.

His other works include:

- "PDCA for ITIL – Metrics, CSFs and workflows for implementing ITIL practices" published by TSO, UK (ISBN 9780117082076) which gives a direction for implementing ITIL processes and designing ITSM solutions.
- "Charm of friendship" published by Pothi, India (ISBN 9789382715924) explaining the importance of friendship for children.
- "Guidance for ITAM – Step by step implementation guide with workflows, metrics, best practices and checklists" published by Servicemanagers.org, Spain (ISBN 9780991320509)
- "Guidance for EAM (Enterprise Asset Management) – Processes, Implementation steps, workflows, metrics, best practices and checklists" published by Servicemanager's.org (ISBN

9780991320516) which details the guidance for planning, developing and improving EAM processes with KPI's, best practices, workflows and checklists.

- "Focus on IAM (Identity and Access Management) – CSF's, metrics, best practices, checklists and guidelines for defining IAM processes and implementing IAM solutions" published by Servicemanager's.org (ISBN 9780991320530).
- "Guidance for Incident Management - According to ISO/IEC 20000 & 9001 Standards, Six Sigma and ITSM Best Practices" published by Servicemanagers.org, Spain (ISBN 9780991320547)
- "Guidance for Problem Management - According to ISO/IEC 20000 & 9001 Standards, Six Sigma and ITSM Best Practices" published by Servicemanagers.org, Spain (ISBN 9780991320554)
- "Guidance for Change Management - According to ISO/IEC 20000 & 9001 Standards, Six Sigma and ITSM Best Practices" published by Servicemanagers.org, Spain (ISBN 9780991320561)
- "Focus on Data Center – An IT consultant's essential guide for working in Data Center environment" published by ServiceManagers.Org (ISBN 9780991320578)
- "Guidance for Service Asset and Configuration Management - According to ISO/IEC 20000 & 9001 Standards, Six Sigma and ITSM Best Practices" published by ServiceManagers.Org (ISBN 9780991320585)

Introduction to Industry4.0

Introduction to Industry 4.0

Industry 4.0 is the fourth industrial revolution which is all about the smart industry, which focuses on providing smart equipment manufacturing (product design, operations, maintenance, quality control and assurance, etc.) and smart processes, which was introduced in 2011 by the German Government.

This section provides introduction to this book with the help of terminology used in Industry 4.0.

Basic Terminology

3D Printing: 3D printing is a process of making three-dimensional solid objects from a digital file. 3D printing enables to produce complex (functional) shapes using less material, than traditional manufacturing methods.

Access: Any information representing the privileges about the identity granting. Access rights for user identities can be categorized as create, read, update, delete, etc.

Additive Manufacturing: A manufacturing process used to create three-dimensional objects in which material is joined and solidified under computer control and other systems.

Advanced Persistent Threats: Network attacks to steal data rather than to cause damage to the network or organization. Here the intruder accesses the network and remains undetected for a long period of time to steal the data from an organization.

Agency: A system that is composed of intelligent agents (hardware and software-based system that has autonomy, social ability and reactivity), such as computers or robots, that cooperate to solve a problem.

Artificial Intelligence: It is the science which uses computers, robots, or other devices to perform functions such as learning, decision making, or other intelligent human behaviors.

Asset: Organization's financial investment or costs on any capability or resource.

Asset Condition: Measure of the health of an asset.

Asset Disposal: Activities or operations necessary to dispose any decommissioned or obsolete asset or a machine in an industry.

Asset Average Life: Average life time of an asset without even a single failure from the date of deployment.

Asset Inspection: Verification and validation performed by the asset handling team to determine the condition of an asset before it is stocked in the inventory.

Asset Identification Label: A label or a tag that will be affixed on an asset like a unique identifier for its identification.

Asset Utilization: Percentage of time an asset is in operation at user or customer premises.

Asset Usage Policy: A policy mentioning the guidelines for end users on usage and protection of assets.

Automated Guided Vehicles: A material handling systems that are battery powered vehicles which are operated, within a particular facility to improve manufacturing and warehouse efficiency.

Best Practices: Practices that are well recognized and which have proved the ability for demonstrating success in respective discipline/industries (e.g. PMBOK for project management, ITIL for IT service management, MOF for IT service management, COBIT for IT governance, etc.).

Beyond Reasonable Repair: Any asset requiring a repair which will cost the same as buying a new asset.

Big Data in Industry 4.0: Data that is generated from various objects in industrial infrastructure (both structured and unstructured) used to gather, process, and enable to make effective decisions in industrial operations.

Biometric Authentication: Authentication system where unique physical characteristics (finger prints, facial features, iris patterns in the eyes, etc.) of user are verified and validated.

Blockchain: Technology enabling the maintenance of a shared digital records in a public or private peer-to-peer network, in a way that it becomes difficult to tamper any individual record. Blockchain technology provides digital trust and automation in the exchange of information between supply chain companies.

Block Diagram: Diagram which depicts the overall hardware architecture of the manufacturing machines/ devices showing the different functions, the requirements they cover and their inter-relationships.

Cabling System: Structured cabling in the industry where the copper and fiber cabling is used as the typical media and are terminated via several types of connectors.

Carbon Footprint: It is the amount of CO2 emissions that are produced in industrial operations.

Change: Any addition or modification or removal of a machine part or component and its attributes in the industrial environment.

Cobot: A machine designed to help the industry staff to do their job more efficiently and safely, with the help of artificial intelligence in an industrial workplace.

Cooling System: Refers to the chillers and air handlers used to regulate ambient temperature and control humidity in the industrial environment. This system might incorporate the air conditioning system used to cool regular building premises.

Chiller: It is a machine that uses chilled water to cool and dehumidify air in industrial production environment.

Compliance: Compliance is a state of being in accordance with established guidelines, specifications, or legislation or the process.

DFSS/ DMADV: A systematic methodology defined in six sigma philosophy, to design products and processes which can meet customer expectations.

Digital Engineering: It is the process of integrating data about an object into the world of digital imagery.

DMAIC: A systematic methodology defined in six sigma philosophy, for continuous process improvement.

Capability: Any intangible assets like processes, knowledge used for managing a machine or a business service in an industry.

Critical Success Factor (CSF): Vital principles necessary for the success of a business or an industry. Based on the CSF's, metrics are developed.

Command: An order to perform a specific activity in the industry which results in a specific outcome.

Core Model: It is a reference model defining the architecture, relationships, attributes which defines the general aspects of a machine or a system.

Cyber Physical System: It is a system where the physical systems (machines) are managed, based on computer-based algorithms.

Data: Any raw facts or figures about industrial processes, machines, operations, etc.

Event: Any occurrence/observation that has significance to the delivery of industrial operations and services. Events are notifications that are received through sensors from different objects to inform if a specific part or a machine or an object's working conditions and issues.

External Customers: Customers who reside outside the manufacturing organization and who purchase the industry's products and services.

Fire Suppression: Devices associated with detecting or extinguishing a fire in the production environment. The most obvious components are water-based sprinklers, gaseous fire suppression systems, and hand-held fire extinguishers.

Functional Escalation: Informing, involving and seeking the help of different technical teams to resolve an issue in a machine or part in a machine.

Goals: Goals are the broad objectives which are abstract and defined for a stipulated time period.

Green Industry: Industry where the mechanical, lighting, electrical and computer systems are designed for minimum energy consumption and minimum environmental impact.

Hierarchical Escalation: Informing, involving, and seeking help from senior levels of management from a specific process or team to assist in escalation.

Horizontal Integration: It is a type of integration where various IT systems are used in the different business planning processes of the manufacturing. It involves an exchange of materials and information both within a company and between several different companies.

Human Machine Interface (HMI): It is the graphical user interface for industrial control and automation, providing a command input and feedback output interface.

Impact: Measure that defines the number of business services affected by a machine or an information system. Usually denoted as 1,2,3,4.

Industrial Augmented Reality: Application of augmented reality in industry manufacturing.

Industrial Device Management: Monitoring system that monitors the health of industrial devices.

Industrial Waste: Anything that the industry no longer requires and therefore discards it by disposing it as cheaply as possible.

IOT Middleware: Software that enables device data to be collated and communicated between services, sensors, software, hardware, etc. giving the flexibility to integrate Industrial equipment enabled with IoT (Internet of Things).

Information: Data that is well organized, processed, which follows a specific structure, and which is easy to understand.

Internal Customers: Users who are from the same manufacturing company, who could be part of the support team or from project teams.

Internet of Services: Any service that can be offered and accessed through internet.

Objectives: Objectives are precise, specific, tangible and measurable accomplishments to be achieved by a business department, specific process, or a team.

Outage: Any unplanned breakdown or reduction in the quality of manufacturing machine or a service delivered in an industry.

Outage Category: A structure that organizes a group of similar types of outages.

Industrial Service Management: Process for managing the lifecycle of industrial manufacturing services to meet the organization's business goals. It defines a standardized process for procurement of assets, receiving (which includes verification and validation), stocking, operations (which is focused on production environment operations), maintenance (which includes repair), retirement and disposal.

Industrial Ethernet: Networking technology used in industrial environment with the help of protocols like ethercat, ethernet/ip, profinet, powerlink, sercos iii, cc-link, modbus/tcp that provide determinism and real-time control.

Industrial Infrastructure: Combination of industrial processes, hardware, software, and people-ware in the industry.

Obsolete Assets: Any asset that no longer serves a purpose or doesn't meet functionality and is out of its warranty period.

Orchestration: The arrangement and coordination of automated tasks, to perform a specific outcome.

Programmable Logic Controller: An industrial digital computer adapted for the control of manufacturing processes, providing relay control, motion control, industrial input and output process control, distributed system, and networking control.

Power Distribution Unit (PDU): This electrical device is used to control the distribution of power to the individual loads, these controls could be as simple as a series of switch or circuit breakers.

Key Performance Indicator (KPI): Vital metrics necessary for an organization to meet its business goals which reflect the CSF's of an organization.

Machine Management Plan: Document which provides the holistic guidance on management of machines and its lifecycle. It provides an overview on risks, assumptions, roles, etc.

Major outage: Any issue that has a huge impact and urgency to the industry stakeholders and disrupts very critical business services.

Metric: Measurements that quantitatively evaluate the performance of industry operations.

Protected Machines: Any machines in industry infrastructure, which is not freely accessible and available for all employees. These machines require special authentication before the employee can use it.

Policy: Policy is a management directive that significantly influences processes and procedures.

Priority: Measure used to identify the importance of a machine or a machine's service. It is generally calculated based on the impact (how many people are affected) and urgency (what is the currency value associated per hour)

Process: A set or sequence of activities that results in achieving an output or business goal.

RFID: Technology that operates based on electromagnetic fields to automatically identify and track tags attached to machines or machinery parts. These tags generally contain electronic information which can be active, passive, or battery-assisted passive types.

Rapid Manufacturing: A new method of manufacturing where industries use 3D printers for small batch custom manufacturing, which will serve as end user products.

Request for Maintenance: Any request on machines for performing the maintenance activities on machines.

Request for Asset Withdrawal: Request from a user or customer to disassociate a machine and return it to the inventory or to the OEM.

Request for Asset Movement: Request to move the machines from one location to another.

Request for Asset Transfer: Request from a user or customer to transfer the machine from one employee to another.

Request for Asset Repair: Request from a user or customer to perform a repair or refurbish a machine.

Resources: Any tangible assets (like machines, people, money, or anything) that will be helpful to deliver an IT service.

Sensors: Hardware components in IoT systems that act like the hands and legs for the industrial automation system that monitors the industrial operational conditions in real time.

Sigma: It is a Greek letter representing a statistical unit of measurement that defines the standard deviation of a population which measures the variability or spread of the data.

Six Sigma: An approach to improve the process quality by eliminating defects and minimizing variability.

Slicing: Dividing a 3D model into hundreds or thousands of horizontal layers which is done with slicing software.

Smart Factory: It is a factory that uses cobots, cloud, IoT, big data and analytics in the factory, to connect everyone and everything together to make better products and give better satisfaction to the employees and the customers.

Smart City: A city that monitors and integrates conditions of all its critical infrastructures using next-generation information technology equipment, using embedded sensors and other equipments to buildings, power grids, roads, transport modes, and other objects in the city.

Smart Product: A smart product that combines the physical and software interfaces (with sensors, IP address, etc.) which has several interactive functions.

Standard: Rules and conventions that help to implement policies and enforce required conventions.

Surge Protector: A device designed to protect against jumps in voltage from the power source.

Swarm Intelligence: Subset of Artificial Intelligence (AI) designed to manage a group of connected machines (to communicate & coordinate) to make industrial operations more automated and smart.

Uninterrupted Power Source (UPS): Uninterruptible Power Supply, or battery backup, is an electrical apparatus that provides emergency power to equipment usually when utility mains fail.

Vertical integration: Strategy used in industries referring the integration of the various IT systems at the different hierarchical levels to deliver an end-to-end solution.

Vision: Vision is a desired goal and intention, meant for long term sustainability of an organization.

Mission: Mission is the statement which defines the status of an organization like 'what is the organization about', 'who are its stakeholders, 'what are its primary activities', etc.

Network Function Virtualization (NFV): It is a technology that transfers hardware-based network functions to software-based applications running on commercial off-the-shelf equipment.

WSN (Wireless Sensors Network): Network of distributed autonomous sensing nodes that use low-power integrated circuits and wireless communication technology to distribute data among the connected sensor devices.

Technology making things smart

With the help of internet, automated software, sensors, and many more new technologies, objects or things have become smart today (they can generate data, collect data, process data and use acquired information to make better decisions).

Examples of smart objects:

A mobile phone was traditionally meant for handling calls and SMS, but now smart phones do many more activities like location detection, voice recognition, face recognition, internet on phone, chat on internet, and many more things.

Doorbell was traditionally meant for ringing the bell, when someone pushes the switch at the entrance of the door. But now smart doorbell can trigger the CCTV camera to take the photo of the visitor and email it to the owner of the house, so that the owner can open the door or reject the request.

Fire alarms was traditionally meant for alarming that there is a fire accident, but smart fire alarms will alarm the owner and can send an email/ give a phone call/ send a SMS to the fire station detecting the address, location details, etc.

Weather and temperature detectors were meant for recording the temperature and the people in houses were supposed to check the temperatures. But now smart temperature detectors can check the temperature and send the temperature and weather reports to the recipient's email addresses and their phone numbers.

How do things become smart?

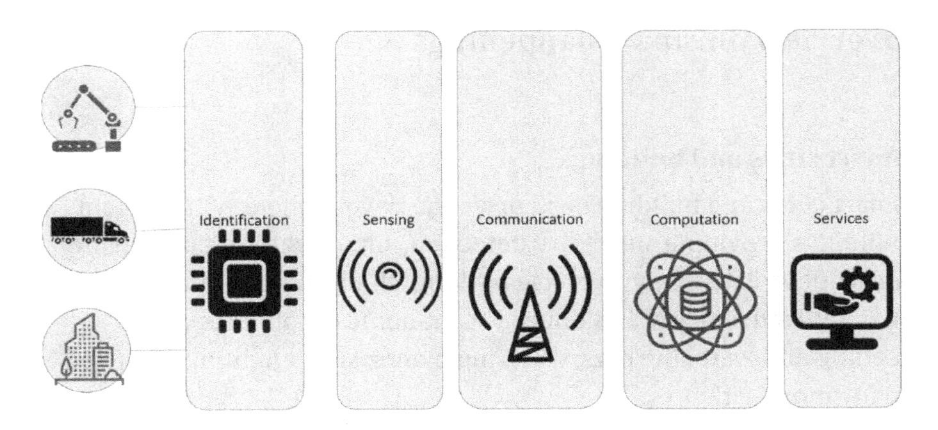

Identification is about things which typically has wireless/ wired sensors and actuators. The sensor/actuator state is the province of operations technology (OT) professionals. HMI (Human machine interface provides the interface for sensing through common communication links to other parts of industrial systems.

Sensing is about sensor data aggregation systems and analog-to-digital data conversion. Sensors will be the integral part of industrial automation systems and provide trigger point and feedback for system control.

Communication is about the IT systems performing pre-processing of the data before it moves on to the data center or cloud. It is possible to have multiple interfaces on the same "Thing", specializing on interfacing with certain properties (AC power, temperature, control). It is possible that multiple interfaced "Things" have the same interface on the M2M end, as long as they use same protocol.

Computation is focused on data analysis, management, and storage on traditional back-end data center systems.

Service is about provision of smart services using all the above-mentioned artefacts.

Overview on smart happenings

Smart cities and buildings

Smart cities and buildings will drive the development of intelligent buildings providing smart fire detection, smart water management, smart physical security, vehicle parking, climate change policies to mandate efficiency in buildings, vacant home capabilities (for feeding and watching pets, watering plants, smart lighting, etc.) and many more activities.

Smart hospitals

Smart hospitals can assist the doctors and nurses to give medication through thorough monitoring which can avoid any kind of ignorance regarding patients. With the help of big data analysis on a patient's historical and current health information, it can identify life critical information and take appropriate actions.

Doctors in smart hospitals can provide guidance to outpatients even remotely with the help of sensors, networking, IoT, etc. to the outpatients.

Devices in smart hospitals can generate alerts and notify when there is an issue or when a device is about to breach the threshold value set for the equipment. For example: When the blood bottle providing blood to a patient is less than 20 ml, it can generate an alert to the nurse to place a new blood bottle for the patient.

Smart inspections in engineering

Underground train tunnel inspection is a perilous job for field engineers, where the technicians will have to enter the tunnel and check the rails, surface, obstacles and make notes of the condition which is a time-consuming, labor intensive, and also has health and safety hazards.

Hence smarter tunnels provide hardware (sensors, digital cameras, etc.) and software technologies that help automate smart inspection tasks to perform the quality checks in the underground train tunnels.

Smart retail

In smart retail, customers have an advantage of knowing the new products, discounts, offers, etc. and they can also avoid long queues, waiting time, etc. which will kill the time unnecessarily.

Smart retail comprises of:

Smart vending machines: They are connected vending machines which will be able to remotely detect and report to inventory.

Smart RFID: It is a RFID tag with a sensor to identify retail products and its associated data.

Digital signage: It changes messages based on the sensor information, directing traffic to various parts of the store.

Smart shelves: It provides intelligent ways like digital price labeling that automatically gives notifications to the merchandisers about the low stock and replenishment.

Self-checkouts: It enables to process their own purchases.

In huge stores like metro, walmart, etc. identifying a product would take some time, but with the help of smart technology (i.e. sensors and RFID technology), customers can identify the goods or products easily. By the help of smart cranes (which can weigh the products and its cartons), retail owners can avoid unnecessary costs in weighing products. These smart cranes can automatically weigh the products, when they are lifting the goods from trucks or ships which saves fuel and also reduces CO2 emissions.

Smart homes

Smart home is a new concept where houses use sensors to monitor daily patterns in home. Sensors installed on doors would track movement regularities, lights would have sensors and would turn on and off based on the human movements like entering or leaving a room. The IoT thermostat improves efficiency of energy use and adjusts the temperature within the house depending on the weather outside.

Likewise, there are many more opportunities, to make the home as a smart home providing benefits to people in home (making life easier) and also being ecofriendly to nature.

Smart waste management

In smart waste management, garbage containers will have sensors which transmits signals to indicate when the threshold limit (70%-80%) reaches. Through the connected networks, garbage containers will send messages to company supervisors for emptying the garbage containers.

Garbage management software can also visualize the capacity of the containers regularly to check if the containers have become full and when it requires a discharge. Accordingly, garbage trucks are sent to those locations where the containers are full.

Smart transportation

Smart transportation provides great number of advantages for both passengers and also the transportation service providers; it can offer scheduling information, ticketing and other services through mobile devices.

In smart transportation, vehicles are embedded with sensors on critical parts to measure and track:

- stress on its parts,

- condition of vehicles,
- air in tyres,
- temperature of the engine and many other parameters.

Accordingly, it communicates this information to gateways and cloud to do the respective analysis.

A smart transportation solution not only deploys sensors to monitor the health vehicles, it also uses CCTV cameras to spot issues on the vehicles. This helps maintenance crews to focus more thoroughly on repairs, providing a safer and more reliable experience for riders.

Smart transportation uses global positioning system (GPS) and processes the data using analytics, accordingly vehicle operators can operate vehicles more efficiently. Smart transportation solutions can also measure the people flow — such as passengers waiting at stations, passengers in each vehicle and many more things.

Analytics can then provide guidance to operators on how to optimize schedules (such as deploying additional vehicles to an overcrowded station), how to improve the comfort of rides and much more. Smart transportation solutions can also get information on the weather and climate, to prepare for and prevent service interruptions.

What is Industry 4.0

Industry 4.0 is the advanced industry which uses modern technology (automation, cloud computing, augmented reality, autonomous robots, simulation software, big data, internet of things, cyber physical systems) to make the industrial operations faster, efficient, effective, and make more effective business decisions.

Fourth industrial revolution is completely different from other industries because its strength is connected products, connected assets, connected infrastructure, connected markets, and connected people which operates based on Big data and analytics, CPS, IoT, Cloud, AI, and many more technologies.

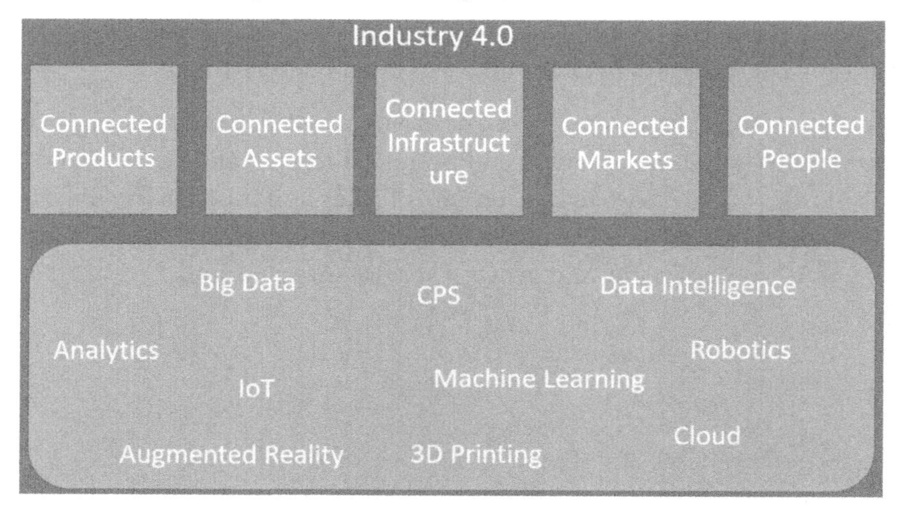

If we understand the history of industries, we will have to start with hydraulics which produced steam engines and was the trigger for first industrial revolution in 1800's. In 1900's there was the second revolution which used electricity in industries and produced products in mass scales. 1950's - 60's, was the beginning for third revolution which used electrical and electronical engineering for production of machines and industrial operations.

Now the fourth industrial revolution is Industry 4.0, which uses combination and aggregation of latest IT technologies, processes, and machinery for faster operations making it more agile and impeccable. Industry 4.0 implements sensors, robotics, and data management technologies to make the machines, infrastructure and the complete industry smarter reducing the time, labor, and making more effective decisions in industrial operations.

Industry 4.0 is driven by four identifiable technology trends:

- networks & connectivity
- data analytics and business intelligence
- digital instructions to the physical world
- human machine interaction through touch interfaces & augmented reality systems

What is so special in Industry 4.0

- Industry 4.0 is based on cloud, big data, IoT, autonomous robotics, etc. Industries would benefit from the latest happenings in technology, improving the efficiency and productivity for manufacturers.
- Industry 4.0 will make the industries more automated as the machines can communicate with different machines enabling better decision making, identifying any issues and eliminating human resources for routine tasks.
- Industry 4.0 will enable the industry to create new business models, and create new products with the help of new concepts like simulation, additive manufacturing, big data, augmented reality, etc.

- Industry 4.0 will give end to end visibility of the industrial operations to the management to make informed decisions about planning strategies, planned production programs and increased supply chain agility.

Concerns in Industry 4.0

- Hesitation to learn new technologies like IoT, big data and advanced analytics. (Generally, in manufacturing industries workforce is used to manual data, spreadsheet-based data management, reporting systems, etc.)
- Adopting automation and laying off human resources.
- Allowing the company sensitive data storage in clouds.
- Managing the technological investment for automation and the knowledge of new IT trends.
- Lack of technological infrastructure and technical resources who have the knowledge on concepts like IoT, cloud, etc.

Benefits of Industry 4.0

- Provides ability to collect, transmit, and interpret data in real time providing better traceability and control.
- Enables tools, machines, and people to communicate more meticulously.
- Can predict failures and trigger maintenance processes autonomously.
- Allows for more flexible methods of getting the right information to the right person at the right time.
- Enhanced productivity through new technology.

- Visibility on end to end real-time data for better decision making, ensuring potential problems can be identified before they occur.
- Effective business continuity through new processes, technologies and monitoring possibilities.
- Better quality products as a result of automation and elimination of human work.

Why Industry 4.0?

Industry 4.0 is the next revolution that brings various solutions that connect factories, components, & stakeholders using smart machines and internet connectivity. Industry 4.0 is the advanced industry which uses modern technology (automation, cloud computing, augmented reality, autonomous robots, simulation software, big data, internet of things, cyber physical systems) to make the industrial operations smart. Here are some reasons why industry 4.0 will bring great advantages to the industries.

- More innovation and better quality

  With the help of latest software tools and technologies, industrial operations would find better ways of designing and creating new products and services.

- Productivity

  Industry 4.0 has better productivity in less time, using fewer raw materials and less energy.

- Digitization

  When industry's operations are digitized, all stakeholders (like retailers, distribution centers, transporters, manufacturers, and suppliers) receive data transparently about the others' supply levels, fulfills orders automatically, and triggers maintenance and upgrades.

- Less dependency on human

  Machines become smarter with better responsive and interactive ability; machines can track their own activity and its results, along with the activity of other products around them. When captured and analyzed, the data generated by these products and services shows how well they are functioning and how they are used.

- Transparency

  With the help of sensors, gateways, cloud and big data in Industry 4.0; it receives real-time, complete, and effective information on every aspect of the smart factory. It can identify visible & invisible issues which might happen due to machine degradation, component wear, etc.

- Reach to customers

  Operations in Industry 4.0 will have better reach to customers based on their needs, demands, tastes, etc. and accordingly products can be developed.

- User friendliness

  With the help of advanced technology (autonomous robots, big data analytics, simulations, software tools, etc.) productivity, maintenance and diagnosis becomes easier.

- Efficiency

  The smart factory has an accurate knowledge of production process, which can help determine the needed materials before production. So that the production and product redundancy can be minimized, over stocking can be avoided, unnecessary motion is reduced, and power consumption can be reduced.

Industry 4.0 in different verticals

Industry 4.0 will change the manufacturing sector, right from purchasing (the raw materials), designing, manufacturing, selling, servicing, etc., every activity in manufacturing would get transformed.

Automotive Industry

With the use of internet technologies consumers can now check the specifications, costs, colors, etc. right from their home and compare with other car manufacturers, car models, etc. Using virtual reality capabilities, automotive companies allows consumers to open doors, sneak a 360-degree peek inside and out view, and even hear authentic sound effects of their potential new models.

Using big data and analytics, IoT, cloud technologies, connected supply chain provides manufacturers with a common platform to operate with real-time visibility, promoting greater interdependency, collaboration, dynamic responsiveness and the flexibility. The connected supply chain drives the costs down, engages the consumers, collects and uses data to serve consumers.

With the use of advanced self-diagnostic systems, automobiles can instantly alert when the tyres need air, when a part must be replaced, when the automobile needs servicing/ replacement. Advanced self-diagnostic systems allow us to proactively service our vehicles, and drastically reduce mechanical failures.

With the use of sensor technologies and IoT, automobiles can monitor data proactively to ensure automobile safety, fuel optimization and even monitoring of cargo. It can also analyse the data gathered from the automobiles to understand the driver preference and their behaviour.

With the technology involved in Industry 4.0, automotive industry is developing self-driving cars using different advanced technologies, collaborating with several objects, software and systems. Self-driving cars are the new and next generation cars using GPS, sensors, cameras for connectivity and algorithms, which can prevent accidents and failures.

Aerospace Industry

With the help of sensors, automated vehicles, IoT, big data and analytics, aerospace industry will improve its productivity, efficiency, and will identify the risks proactively and also resolve the issues reactively.

As aerospace manufacturing units have huge equipments, AGV's, robots and cobots work alongside humans in assembly lines. These connected technologies increase productivity and decrease defect rates.

Aero-equipments are fitted with sensors which can identify when equipment is wrongly or loosely fixed. Not only in manufacturing process, but also when the machines are flying they generate data. Based on the real-time data received from equipments and sensors (while flying), supervisors can adjust the way aircraft is flown and take care of potential issues before they become real problems that end up grounding airplanes.

Sensors monitor a wide range of vital parameters throughout the systems and sub systems of an aircraft and can log all the issues during the travel and will provide detailed maintenance information once the plane is grounded.

With the help of big data and analytics, data will be gathered from the previous flights & its issues, to make better decisions in manufacturing new models with respect to things like the fuel consumption rate, stress levels, temperature, and power usage.

Supply chain Industry

Industry 4.0 will bring tremendous changes in supply chain management explicitly to the warehouse, transport logistics, procurement, and order fulfilment processes.

Technologies, such as virtual and augmented realities, artificial intelligence, 3D-printing and simulation, will all result in great opportunities. On the other hand, big data analytics, cloud technology, cybersecurity, IoT, miniaturization of electronics, AIDC, RFID, robotics, drones, nanotechnology, M2M communication and BI will bring plethora of opportunities in the supply chain industry.

Virtual and augmented reality will be used for equipment repair, safety, quality monitoring, and order picking.

Artificial intelligence will determine optimal inventory levels, to eliminate repetitive purchasing tasks involved in the make-or-buy decision process, to forecast the supply-and-demand information.

Robotics will be used in materials handling and assembly.

The most relevant benefits for supply chain industry are increased flexibility, quality standards, efficiency and productivity. It will enable mass customization, allowing companies to meet customers' demands, creating value through constantly introducing new products and services to the market.

Principles for Industry 4.0

Principles for Industry 4.0 can be derived from the Industry 4.0 components like autonomous robots, cyber physical systems, cloud computing, additive manufacturing, internet of things, etc.

Interoperability: Interoperability is a very important principle for Industry 4.0. Interoperability in Industry 4.0 is evident by the CPS's in the plant, which can communicate with objects, machines and people.

Virtualization: Virtualization in Industry 4.0 enables CPS systems to simulate and create a virtual copy of the physical machines. In simple words, there will be a virtual copy of everything.

Decentralization: Decentralization in Industry 4.0 enables the CPS's to work independently. For example, numerous cobots in Industry 4.0 cannot be controlled by central server, instead cobots are embedded with computers to act own their own creating a more flexible environment for production. Only in the cases of failures or having conflicting goals, the issue is delegated to a higher level.

Real-Time Capability: Real time capability in Industry 4.0 enables to collect real time data, store, analyze, and make decisions according to new findings. This capability helps to understand the internal processes in a more detailed way such as the failure of a machine in production line.

Service-Orientation: Smart factories are built on service-oriented architecture enabling people and smart objects/devices to connect efficiently through the Internet of Services to create products based on the customer's specifications.

Modularity: Modularity in Industry 4.0 enables to adapt swiftly and smoothly to seasonal changes and market trends. These modules can be added using the Plug & Play principle. Modular systems can flexibly adapt to changing requirements by replacing or expanding individual modules. Therefore, modular systems can be easily adjusted in case of seasonal fluctuations or changed product characteristics.

Transparency: Transparency in Industry 4.0 enables those information systems to create virtual copies of the physical world by configuration of digital data into sensor data. For this to be achieved, raw sensor data has to be aggregated with compatible context data.

Proactivity: Proactivity in Industry 4.0 will enable workforce and systems to predict, foresee and act before any issues occur. This ability of the smart factory to predict future outcomes is based on historical and real-time data. Proactivity in smart factory will improve uptime, yield, quality, and will also avoid safety issues.

Best examples could be identifying anomalies, restocking and replenishing inventory, predictively addressing quality issues, and monitoring safety and maintenance concerns.

Technical assistance: Technical assistance in Industry 4.0 enables the cyber-enabled systems to physically support human resources by handling various tasks, which are considered time-consuming, harmful and exhausting to people. Technical assistance capability supports humans through comprehensive aggregation and visualization of information for better decision-making and quick solutions for issues in Industry 4.0.

Smart Industry Architecture

The architecture for smart factory is built up of many devices like sensors, RFID's, PLC's, smart meters, Ethernet switches, Human machine interfaces, etc. which reside at four different layers of the smart industry representing an engineering system which runs based on interconnection, collaboration and execution.

Four layers of smart industry can be mentioned as:

1. Physical layer
2. Network layer
3. Cloud application layer
4. Terminal layer

Physical layer: This layer comprises all intelligent manufacturing equipment, sensors, RFID's, PLC's, conveyor equipment's, and packing products. It is responsible for execution of tasks such as processing, monitoring, assembling, etc.

Network layer: This layer comprises networking technologies and topologies (field bus, Modbus, and EtherCAT) to connect the four layers.

Cloud application layer: This layer performs the cloud storage activities like storage, knowledge management, QoS management, ontology modeling, information evaluation, etc. The manufacturing data is uploaded to the cloud platform to form a semantic data model. Based on the knowledge base, different tools are used to reason the equipment operating mode. This layer provides supports for the fault alarm, the resource allocation, analytics and the scheduling optimization.

Terminal layer: This layer is used to visualize the results of cloud processing and access the transactions happening in the industry (including customers). This layer comprises end-user devices (PDA's, computers, etc.).

RAMI 4.0

There is framework called RAMI 4.0 (Reference Architectural Model for Industry 4.0) which defines the architecture in a three-dimensional map showing Industry 4.0 in a structured manner. This model follows service-oriented architecture, combining all elements and IT components in a six layered and lifecycle model.

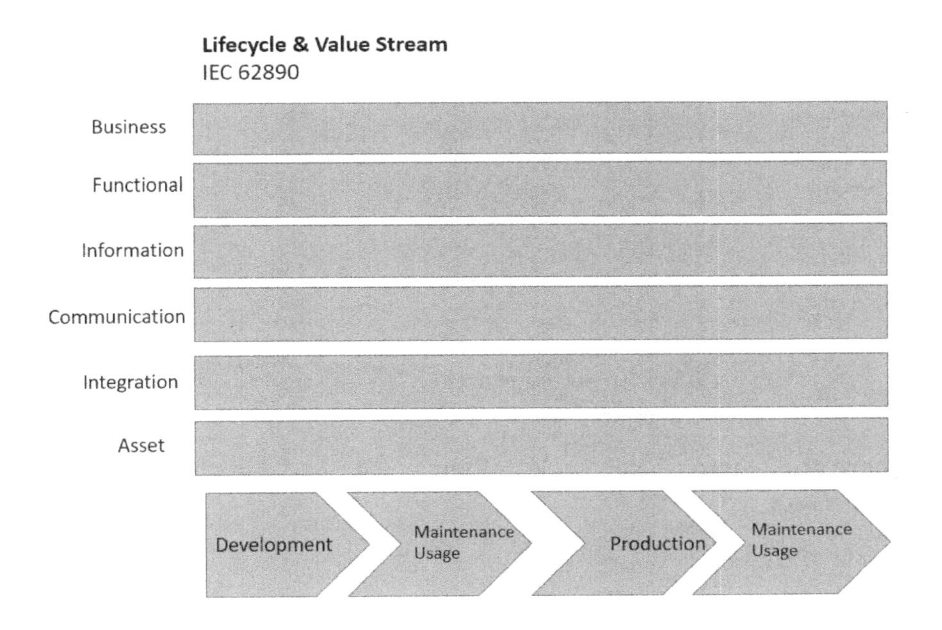

Six layers defined in RAMI model are:

1. Assets

2. Integration
3. Communication
4. Information
5. Functional
6. Business

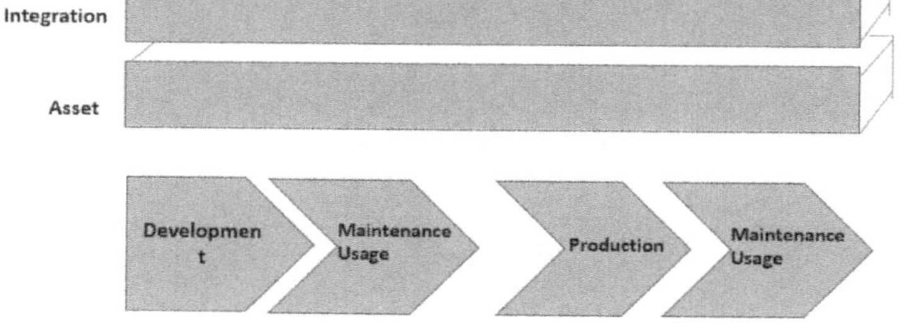

There are 7 hierarchical levels defined in RAMI 4.0 which can be mentioned as:

1. Product
2. Field device
3. Control device
4. Station
5. Work centers

6. Enterprise Connected world

Concise view into Industry 4.0 components & technologies

Concise view into Industry 4.0

Some of the major concepts which will drive Industry 4.0 are:

1) Autonomous robots

2) Simulation

3) Cloud computing

4) Augmented reality

5) Cybersecurity

6) Additive manufacturing

7) Bigdata and analytics

8) Internet of things

Sensors

Industry 4.0 will use various kinds of sensors collecting data, pushing it, sharing it with gateways and cloud storage for analyzing data and to make effective decisions.

Sensors can be classified into different types; some prominent sensors can be mentioned as:

Temperature sensors

Temperature sensors are used to measure amount of heat energy that allows to detect a physical change in temperature from a particular source, accordingly it converts the data for a device or user. These sensors can be used in manufacturing A/C control machines, air coolers, fans, refrigerators, ovens, washing machines, medical equipments, agricultural equipments, etc. and similar devices.

Temperature Sensors are further classified based on its functionality as:

- Thermocouple sensors which detects temperature with change in voltage.
- Resistor temperature sensors which detects the temperature or resistance going up.
- Infrared sensors which detects temperature by intercepting a portion of emitted infrared energy of the object or substance and sensing its intensity.
- Proximity sensor detects the presence or absence of a nearby object, or properties of that object, and converts it into signal which can be easily read by user or a simple electronic instrument without getting in contact with them.

Pressure sensor

A pressure sensor is a device that senses pressure and converts it into an electric signal. These sensors make it possible to monitor systems and devices that are pressure propelled. Based on the deviation in standard pressure range, the devices notify the system administrator about any problems that should be fixed. Deployment of these sensors is useful in manufacturing whole water and heating systems, as it is easy to detect any fluctuation or drops in pressure.

Pressure sensors are further classified into different types as:

- Water quality sensor: Water quality sensors are used in water distribution systems to detect the water quality and Ion monitoring.
- Chlorine Residual Sensor: It measures chlorine residual (i.e. free chlorine, monochloramine & total chlorine) in water and is most widely used disinfectant because of its efficiency and cost.
- Total Organic Carbon Sensor: TOC sensor is used to measure organic elements in water.
- Turbidity Sensor: Turbidity sensors measure suspended solids in water, typically it is used in river and stream gaging, wastewater and effluent measurement.
- Conductivity Sensor: Conductivity measurements are carried out in industrial processes primarily to get information on total ionic concentrations (i.e. dissolved compounds) in water solutions.
- pH Sensor: It is used to measure the pH level in the dissolved water, which shows the acidic or alkaline nature.

Chemical sensor

Chemical sensors are devices used to track & indicate changes in liquid or air chemical changes.

Main use cases of chemical sensors can be found in Industrial environmental monitoring and process control, for detecting intentionally or accidentally released chemicals, explosive and radioactive materials.

Gas sensor

Gas sensors are used to monitor changes of the air quality and detect the presence of various gases, they are used in many industries such as manufacturing, agriculture and health. These sensors will help in air quality monitoring, detection of toxic or combustible gas, hazardous gas monitoring in coal mines, oil & gas industries, chemical laboratory research, manufacturing – paints, plastics, rubber, pharmaceutical & petrochemical etc.

Some of the common gas sensors are carbon dioxide sensor, carbon monoxide sensor, hydrogen sensor, air pollution sensor, nitrogen oxide sensor, oxygen sensor, etc.

Smoke sensor

A smoke sensor is a device that senses smoke (airborne particulates & gases), it's level and notifies the user about any problem that occurs in different industries.

Smoke sensors are extensively used by manufacturing industry, HVAC, buildings and accommodation infra to detect fire and gas incidences. This serves to protect people working in dangerous environments, as the whole system is much more effective compared to the older ones.

IR sensors

An infrared sensor is a sensor which detects certain characteristics of its surroundings by either emitting or detecting infrared radiation.

IR sensors are used in:

- Healthcare equipments as they can monitor aspects like blood flow, blood pressure, etc.
- Regular smart devices such as smartwatches and smartphones.
- Home appliances & remote control, wearable electronics, optical communication, high-level security in your home.

Motion detection sensors

Motion detector sensors detect the physical movement in an area and transforms motion into an electric signal.

Motion detection sensors are very useful in the security industry, in the form of intrusion detection systems, automatics door control, boom barrier, smart camera (i.e. motion based capture/video recording), toll plaza, automatic parking systems, automated sinks/toilet flusher, hand dryers, energy management systems (i.e. Automated lighting, AC, Fan, Appliances control), etc.

Gyroscope sensors

Gyroscope sensors are used to measure the angular rate which is defined as speed of rotation around an axis.

These sensors are generally used in car navigation systems, game controllers, cellular & camera devices, consumer electronics, robotics control, drone & RC (Radio Controlled) helicopter or UAV control, Vehicle control/ADAS, etc.

In this chapter we have studied about sensors, its importance in Industry 4.0, different types of sensors and its applications and with this topics we will conclude the topic on sensors.

Autonomous robots

Autonomous robots play a very important role in Industry 4.0, smart factories or smart industries, would need intelligent environments and one of the component or artifact in this intelligent environment is autonomous robots which generally works based on deliberation, the set of behaviors and through the IP communication.

Before we go ahead further about "Autonomous robots", lets understand the etymology of the words 'Autonomous' & 'Robot'. The word autonomous is derived from the Greek word 'autonomous' which means having one's own laws and the word robot is derived from the Czech word 'robota' which means 'a worker of forced labor'.

How do automated robots make decisions?

Well, nothing works as a magic or by chance, for every activity there is an input, trigger and output.

Any autonomous robot would perform actions based on the agreed conditions, as mentioned below:

- when all the conditions are true.
- when some specific conditions are true.
- as per the requirements of the customer which will be designed by the robot manufacturer.

The development of autonomous robots involves several subtasks like:

- Modeling and prototyping the robot mechanism understanding kinematics, dynamics, and odometry.
- Reliable control of the actuators.
- Generation of task-specific motions.

- Sensor selection and integration of sensors.
- Coping with noise and uncertainty.
- Creation of flexible control policies.

Human interaction with autonomous robots

In earlier days, interaction with autonomous robots used to happen with graphical programming interfaces.

In Industry 4.0, interaction with autonomous robots will happen through graphical programming interfaces, deictic interfaces, voice recognition and also through IP addresses (every robot will have an IP address which will be treated like a node or client and there would be a server which would be communicating, commanding and instructing what actions have to be performed).

What are autonomous robots made of?

These autonomous robots have:

- Processors
- Operating system
- Sensors
- Kickers
- Actuators
- NIC cards
- Video camera and frame grabbers
- And many more based on the scope of work of a robot

Autonomous robots in Industry 4.0 can be used

- For logistics tasks

- For emergency and disaster management
- For Industrial cleaning
- For watering machines

Where else these automated robots can be used?

Automated robots can be used in:

- House cleaning
- Lawn mowing
- Services for the disabled and assistance to the elderly people
- Vision systems
- Planetary explosion
- Mine site cleaning
- And many more

So now we have understood what are autonomous robots, how they make decisions, its components, its applications and with this topic, we complete the holistic information on autonomous robots.

Simulation

Simulations provide the schematic representations of the goods or services. They are also helpful to forecast and estimate the dynamic operations of the industry with respect to seasons, market conditions, geographical locations, etc.

Simulation in Industry 4.0 will be very much essential as simulation software will be used to draft, design, test, simulate, and develop virtual components in Industry 4.0 infrastructure and also to help the day-to-day operations in Industry 4.0.

The etymology of the word 'Simulation' comes from the Latin language which means 'to make like or represent'. Simulation defines the representation of the system, its properties, relationships, etc., which serves as a plan from which a final product can be defined. Simulations are also made based on the history of the operations and drawing inferences from it.

Simulation software in Industry 4.0 would provide detailed views in 360° which can give a precise picture of the manufacturing components, gives meticulous information to test a huge number of possible configurations, to validate the designs, to optimize the space, to optimize the costs, to reduce energy consumption and carbon footprint ensuring the industry is following the green industrial practices.

Simulations in Industry 4.0 can also give an idea on demands based on the patterns of business activities with respect to historical information (during months/ seasons, etc.) and will provide the right numbers in supplying the raw materials with no surplus nor shortage. It will also reduce the unnecessary costs in stocks (raw materials), unnecessary stocking of goods, which will reduce the operational expenditure (OPEX) and total cost of ownership.

Simulations and virtual simulators in Industry 4.0 can also reduce the number of demo machines and costs associated with it. They would give better interface to understand the products or machines, gives training on developed products and provides virtual environment for trial-and-error kind of experience.

One of the software type used for simulations in Industry 4.0 is Discrete Event Simulation (DES) software which will be used to model every activity in the manufacturing process for overall operations, demand, supply, availability, capacity, changes, installations, testing and improvement activities.

Discrete Event Simulation software will create models and prototypes for the new machines or services, which will enable to understand the pros and cons of designs, do thorough testing on the industrial products, get approval from the management and accordingly release or launch the products or services. DES software will also be of great aid for existing products and services, it can show the relationships between different components in industry showing how services or machines will be affected and the impact on production. Accordingly, revised courses of action can then be assessed and implemented.

In this chapter we have realized the importance of simulation in Industry 4.0 and with this topic, we complete the bird's eye view on simulation.

Cyber-physical systems

Cyber-Physical Systems (CPS) are physical systems controlled by computers and internet which involves digital, analog, physical, and human components engineered for providing different functionalities through integrated physics and logic.

Examples of Cyber-physical systems can be mentioned as a self-driving car, Segway scooters, smart spoons, smart robots, an intelligent transportation system, etc.

CPS is the aggregation of computing and communication with physical processes, focusing on prime benefits like safer and efficient systems.

Working model of cyber-physical systems

Working model of Cyber-physical systems can be defined in six stages as identification, integration, monitoring, networking, storage and computing.

Identification: Unique identification is necessary in industry manufacturing, by which a machine can communicate with other machine/s and users. This identification of machines or parts is done through RFID technology.

Integration: Sensors and actuators are integrated for the movement of machines, identifying, monitoring changes, and communicating with other machines in the environment.

Monitoring: Monitoring of physical processes and environment is a fundamental function of CPS. It is also used to give feedback on any past actions which are taken by the CPS and ensure correct operations in the future.

Networking: This step deals with the data aggregation and diffusion.

Storage: Data generated by the networked sensors and machines are stored and analyzed for a better decision making.

CPS in vertical integration

CPS systems are used in vertical integration to enable factories to react rapidly to the changes in demand or stock levels and faults. With the use of CPS systems, smart factories can get the advantage of real-time quality, time, resources and costs in comparison with traditional industries.

These advantages are possible through the flexible network of CPS-based systems which automatically oversee production processes. With the CPS based vertical integration, the production process can achieve a dynamic reallocation of production schedule according to discrepancies in prices, amendments to orders, fluctuations in quality, etc. which optimizes the process structure and makes the production process more flexible.

CPS in horizontal integration

CPS systems are used in horizontal integration which provides networking creating transparency and flexibility across the entire process chains from purchasing through production to sales and end user satisfaction.

In this chapter we have studied what is CPS, working model of CPS, and with this, we complete the holistic information in CPS.

Cloud computing

From the last few years, cloud computing has become an indispensable value for all industries because of its cost efficiency, flexibility, eco-friendliness, user-friendliness, and reduced maintenance work on managing the IT infrastructure. Cloud computing in Industry 4.0 will provide great value because of its unprecedented computing, storage and networking capabilities.

Before we go ahead about cloud computing, lets first understand what is cloud computing?

Cloud Computing in simple language is the technology that provides on demand service, without the ownership of risks, fixed costs, maintenance costs on its underlying hardware components, software components, and its associated licenses. A user of cloud computing technology would just need a computer provisioned with internet access.

Some definitions on cloud computing, in my opinion:

- Practice that uses the distributed computing science on internet or intranet to run, store, and manage the data.
- Technology that provides on demand services (storage, access, provision for software development, provision of software, provision of infrastructure, etc.) over the internet.
- Provision of different IT services through latent infrastructure, using the concepts of shared services, converged infrastructure and virtualization.
- The implementation of managed services with the help of latent infrastructure, using the concepts of shared services, converged infrastructure and virtualization.

Industry 4.0 will have to rely on cloud computing majorly:

- As it will have many machines which will be networked globally involving various locations, employees, partners, etc. so it is very obvious that the data will explode and is very obvious that it will travel across countries and continents.
- As it can be used for storing CRM data, corporate email data and file storage.
- As it can optimize business processes, enabling a more efficient supply chain and provide predictive maintenance.
- As it can support elastic scaling and hyper scaling of new services.
- As it will provide the backbone for storing the data and providing the channel to make the industry data accessible, available, and secure.

Hence cloud computing can be considered as one of the core enablers for Industry 4.0 and with this topic we complete the holistic information in cloud computing.

Augmented reality

While developing complex, rare, and unique tasks in industries, it is quite obvious that staff might make mistakes because of bad memory or misunderstandings. Hence augmented reality becomes the solution for these kinds of issues.

Augmented reality uses rich visualizations, that you will be able to see, understand every part and its functionality through a tablet/ smart phone/ smart glasses.

Augmented reality (AR) is a field of computer research which deals with the combination of real-world and computer-generated data. AR uses video imagery which is digitally processed and augments by adding computer-generated graphics. The basic idea of augmented reality is to superimpose graphics, audio and other sense enhancements over a real-world environment in real-time.

Augmented Reality can be used in proactive and reactive maintenance activities as a guide / simulation/ troubleshooting guide with step by step procedures to perform complex maintenance procedures.

Proactive and reactive maintenance on machines requires greater engineering skills, which would require implementation of augmented reality solutions. AR will also be helpful in the process of training, providing digital visualizations to show users exactly where to put parts, what parts may be broken and also give information on the part. As AR uses a digital overlay onto the real world, the abstract nature of text-based instructions is removed. AR will help operators to perform maintenance tasks which can be carried out by the machine operators without a dedicated intervention from the maintenance department. Simple maintenance tasks can be represented as operations like cleaning, lubrication and inspection.

Augmented reality in different phases of industrial manufacturing

If industrial manufacturing is divided into phases like design & prototyping, assembling, inspection, end-to-end assembling, packaging and training. AR will be helpful in every phase:

Design and prototyping: Augmented reality provides tools that improve design, prototyping and visualization in the design phase. It can also detect the design flaws and also test alternatives and visualizes simulation results. It will also avoid the need of physical prototypes.

Assembling: AR can make big difference to installation and assembly, making manufacturing really smart by providing guidance on constructing parts and performing complex tasks.

Inspection: By using tablets/ smart glasses/ smart phones with the combination of AR, manufactured parts and machines can be inspected without going to the location. AR supports in quality control processes allowing to check whether the items produced are aligning to the manufacturing standards.

End to end assembling: It will provide virtual scenarios of the end to end machine construction.

Packaging: It will provide an idea what kind of machines would carry huge machines. AR tools can improve efficiency of warehouse management operations and logistics supporting operators during indoor navigation and picking operations.

Training: AR guided training will be very helpful for experienced people and also new technicians at the beginning of the learning curve.

Also, AR will enable the interface to manage risk and safety of operators and equipment working in the facilities.

In this chapter we have studied what is AR, its lifecycle phases and with this we complete the overview on augmented reality.

Cybersecurity

Cybersecurity refers to the security offered to online services and information systems to protect the information communicating between networks.

As Industry 4.0 runs on the interconnection of machines, sensors, gateways, middleware which uses cloud storage with big data and analytics concepts. Main concerns would be the attacks perpetrated against their availability, denial of service, confidentiality problems, integrity problems, and authentication issues.

Hence Cybersecurity in Industry 4.0 would:

- Ensure providing confidentiality, integrity and availability for digital supply network systems.
- Provide restricted access to operations on physical systems.
- Detect malware, abnormal and suspicious behavior.
- Prevent Cyber-attacks to CPS.
- Protect from advanced persistent threats, Cyber espionage, and Cyber terrorism.

    Advanced persistent threats: It is a targeted attack which uses advanced techniques and will remain in the information systems for long period with objective to steal data. There are 6 stages of APT which can be mentioned as: Reconnaissance, Delivery, Exploitation, Operation, Data collection and Exfiltration.

    Cyber espionage: It is the internet attack on a network that enables an unauthorized user or users access and spy the data.

    Cyber terrorism: The attack on information systems using internet.

Cyber security should become an integral part of the strategy, design, and operations of Industry 4.0 as the threat agents could be:

- Employees,
- Competitors,
- Hackers
- Cyber criminals & terrorists

And the threats could be:

- Physical damage
- Loss of information
- Information leakage
- Code injection
- Worms / Trojans
- Botnets

Hence Industry 4.0 should have a defined cyber security process managing and mitigating the cybersecurity risks as mentioned below:

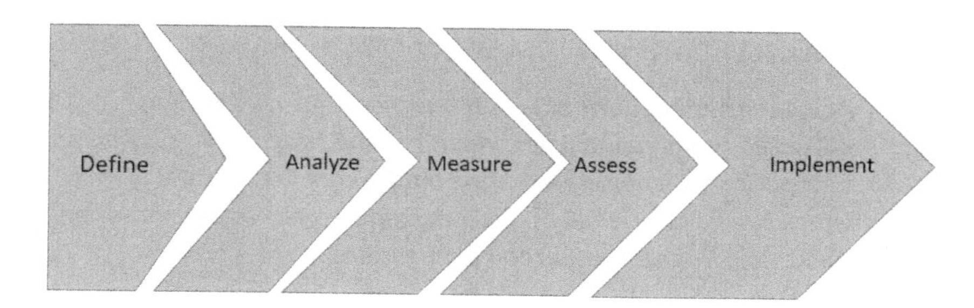

Define phase involves –

- defining the scope and priorities
- selection of critical assets

Analyze phase involves –

- analysis of the threats by performing vulnerability assessments

Measure phase involves –

- measuring potential impact through monetary costs

Assess phase involves –

- assessing the options
- developing potential solutions
- identifying residual risks

Implement phase involves –

- implementing solutions
- conducting trainings and awareness campaigns
- defining and updating cyber security management system

In this chapter we have understood what is cybersecurity, its applications, its process lifecycle and with this, we complete the holistic information on cybersecurity.

Additive manufacturing

Additive manufacturing will have great importance in Industry 4.0 as it can produce the desired components faster, flexibly and precisely which can turn data into components and products at an incredible speed. It involves less prototype construction, fewer dies, less post-processing.

The procedure for 3D printing still follows the same traditional ink-jet printing but the small difference is that the ink-jet printers apply ink to paper and 3D printers inject materials in successive patterns to build a three-dimensional object. 3D printing covers a range of printing technologies that apply different approaches.

There are various technologies used in 3D printers. They are broadly based on the following technologies:

- **Material Extrusion** based 3D printers develop object layers through a semi-liquid material from a print head nozzle.
- **Photo Polymer** works by selectively solidifying a liquid resin called as a photopolymer which hardens when exposed to a laser or any other light source.
- **Selective Deposition Lamination** develops successive layers of cut paper, metal or plastic that are stuck together to build up a solid object. They can also be used to print in 3D color.
- **Binding 3D printing** technology creates objects layers by selectively amalgamating the granules of a very fine powder. Such 'granular materials binding' can be achieved by jetting an adhesive onto successive powder layers, or by fusing powder granules together using a laser or other heat source.

Printers can be classified into the below mentioned types:

- Material extrusion

- Vat Photopolymerization, these printers are further classified into Stereolithography (SLA) Printers, Continuous Liquid Interface Production (CLIP), and Digital Light Processing (DLP).
- Material Jetting Printers
- Powder Bed Fusion Printers
- Binder Jetting
- Direct Energy Deposition
- Laminated object manufacturing

While these technologies differ significantly, they all pursue the same purpose that is to create complex and customizable designs on an individual basis. This enables companies to design and produce new products, easily individualizing them to meet the end-user's growing needs for customization and personalization.

Initially 3D printing made objects from plastic, which were used as visual models. As the mechanical properties and the temperature stability of these objects is limited, now the advancement is in the production of lasting and functional components with metals and ceramics (which provide the defined mechanical and thermal properties). So now, 3D printing produces not only models but real, functioning components with sufficient mechanical properties and adequate heat resistance. Currently, 3D printing uses two methods that is with the help of metal powder and laser beams.

3D printing is a success in Industry 4.0 as it can offer:

- rapid prototyping with low cost and in less time
- full-fledged testing on the prototypes, before the real production is started
- chances to understand the product from 360 degrees, to bring more quality and innovation

- waste reduction

3D printing holistic process

Holistic view of 3D printing process can be defined through the below mentioned steps:

- Firstly, creating images using traditional CAD software or by scanning the set of 3D images.
- 3D models are created from the bottom up approach with 3D modeling CAD software or based on data generated with a 3D scanner.
- Slicing software will be used to divide a 3D model into hundreds or thousands of horizontal layers.
- Selection of the materials (which includes plastics, ceramics, resins, metals, sand, textiles, biomaterials, glass, food and even lunar dust) that will best achieve the specific properties required for your object and will enable you to achieve the precise design.
- Once the 3D model is sliced, it is sent into the 3D printer for the process of printing. Once the file is uploaded in a 3D printer, the object is ready to be 3D printed layer by layer.

Building the product using 3D printing involves many more activities like:

1. Assembly modeling focusing on creating, editing assembly mates, layouts, and best practices for managing large assemblies.
2. Sheet metal building focusing on building complex sheet metal parts and converting solid bodies directly into sheet metals.
3. Modeling validation is about verifying & validating and troubleshooting topology errors (if necessary) and repairing imported geometry.
4. Post processing which focuses de-binding and sintering.

- Finally, when the object is first printed; it cannot be directly used or delivered until it has been sanded, lacquered or painted to complete it as intended.

Additive manufacturing in different domains

Automotive: 3D printing will be helpful in designing parts, spare parts, tools, interior elements; automotive engineers can use printed parts to restore old cars.

Aviation: 3D Printing is used in designing the parts of airplanes (like light weight frames, printed titanium parts), and will also help the aeronautical engineers to rethink the basics of airplanes. 3D printing will also allow companies to produce light-weight aircraft with better aerodynamics.

Aerospace: 3D printing will enable the aerospace industry to design complex parts and produce 4D programmable materials that react to conditions in very specific ways.

Construction: 3D printing will be important in construction businesses for designing the prototypes of houses and buildings, and it is already producing great results. 3D printing will also cut down manpower and time in visualizing designs for clients.

In this chapter we have understood what is additive manufacturing, its importance, 3D printing process, its application in different domains and with this topic, we will complete the holistic information in additive manufacturing.

Big Data

Big Data defines a process for analysis on large volumes of data, which is collected, and with the help of AI it is analyzed to make better decisions in the organization.

Artificial Intelligence (AI) is one of the most important artifact which will make machines and factories smart. AI in Industry 4.0 encompasses technologies (machine learning, deep learning, quantum computing, etc.) which enables the machines, networking devices, chips, sensors and the software programs to become intelligent (using logic) making predictions, recognizing patterns, and taking actions.

An example of AI technology in Industry 4.0 would be: autonomous vehicles and robots (which will have embedded motion cameras) can detect if there are any obstacles, it will apply brakes, it can change the direction of the vehicle and move towards the destination.

So bigdata in Industry 4.0 will play a phenomenal role as it is going to access and process humungous volumes of data which is structured, unstructured, and semi-structured data.

In the Industry 4.0, data is generated by each machine, by each interaction between machines, by each sensor from each machine, controlling systems, engineers, and so on. All this data is saved in the cloud in order to identify patterns and make decisions from it. Availability of the right data from all aspects of product development, production and testing from big data adds a new dimension to manufacturing operations, enabling targeted innovation, marketing and decision-making.

One simple example would be, data from the day-to-day operations is captured every second and sent to cloud environment. The data saved in the cloud would then be analyzed using sophisticated statistical tools, and a concrete action would be taken to change costing, quality, scope requirements based on knowledge gained from the analysis.

Overview of Big Data Lifecycle phases

Overview of Big Data Lifecycle in Industry 4.0 can be described as:

- **Business requirements understanding**: Understanding the needs clearly, what the business objectives, what is the scope in Industry 4.0, and desired results of the project. As many different forms of data is gathered, you need to understand what data is prioritized one.
- **Data identification**: This stage determines the origin of data (collected from machines, cameras, people, etc.) and also to determine the reliability of the data collected.
- **Data acquisition and filtering**: In this stage, the data is gathered from different sources. After the acquisition, a first step of filtering is conducted to filter out corrupt data. Additionally, data that is not necessary for the analysis will be filtered out as well. The filtering step will be applied on each data source individually, and the data is sent into the cloud.
- **Data extraction**: Data identified in the two previous stages may be incompatible with the data analytics tools that will perform the actual analysis. In order to deal with this problem, the data extraction stage focuses to extract data from different formats from data sources and transforming these into a format that the Big Data tool can process and analyze.
- **Data aggregation and representation**: The data aggregation and representation stage is focused to integrate multiple datasets to arrive at a unified view.

- **Data analysis**: This stage focuses to carry out the actual analysis task which involves executing the algorithm that makes the computations and will lead to the actual result. Data Analysis is simple and could also be a complex task, depending on the required analysis type. This analysis produces the 'actual value' of the Big Data science for Industry 4.0. (Note: If all previous stages have been executed carefully, the results will be factual and correct.)
- **Data visualization**: This stage is involved in using data visualization techniques and tools to graphically communicate the analysis results for effective interpretation by business users.
- **Utilization of analysis**: Final step of the life-cycle is to use the results in industry.

Examples of Big data applications in Industry 4.0

Power consumption: As big data integrates, analyzes, and uses real-time information from various sources, it can analyze data on power utilization habits of different departments and machines in industry, by using smart meters. Analysis from big data can help decision makers come up with wise decisions in terms of predicting the need of the power supply in the future.

Robots & Automated guided vehicles: Big data analyzes the data from autonomous robots and automated guided vehicles which can help improve parts and machine movement to minimize unnecessary movement, providing alternative routes and reducing the number of accidents by analyzing the history of mishaps, including factors such as their cause and the speed.

Moreover, the big data collected from AGV's and autonomous robots can help consolidate shipments and optimize shipping movements by reducing supply chain wastage. Smart transport data can also provide many benefits, such as reducing the environmental impact and increase safety and improving end-to-end user experience, among many others.

Policy definition: Big data analytics can also help the management establish and implement satisfactory policies in Industry 4.0, by understanding the needs of the people, machines in terms of predictive and reactive maintenance, etc.

Big data analytics provides many advantages like:

- Real-time system health monitoring
- Predictive analytics solutions
- Intelligent scheduling and forecasting models
- Root cause analysis for complex and inter-dependent system failures

In this chapter we have understood what is Bigdata, its lifecycle, examples of bigdata, its importance and with this, we complete the basic overview in Big Data.

MES

Traditionally MES is an information system which liaised between ERP systems, individual machines and automation controls. This system provides overall control and management of the factory floor and provides updated information to the ERP.

In Industry 4.0, data is generated from numerous objects (smart materials, machines, devices, etc.) hence we need a system:

- To manage and monitor the work in progress guaranteeing quality, productivity, costs, and ensure work is delivered as per the defined timelines.
- To consolidate data and send (from the factory) IIoT data to other systems.
- To perform advanced analysis inside the smart factory.
- Consolidate, correlate, transmit, and store IIoT data.
- To provide brokering connections between autonomous products and equipment.
- To plan the optimal production process.
- To support in decision making.
- To document the production process.
- To monitor and ensure product quality, process ability, material and resource availability.
- To process IoT data including events, locations and integrate with augmented reality.

And the solution is modern MES (Manufacturing Execution System) systems which can integrate two sides of the smart factory (i.e. enterprise resource planning (ERP) and traditional manufacturing execution systems).

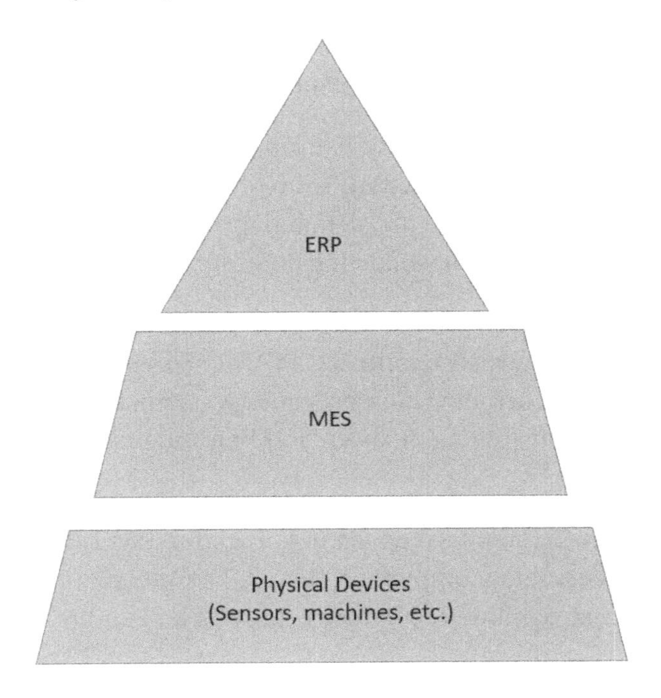

Processes involved in MES

MES (Manufacturing Execution Systems) is a consolidated information system for manufacturing industry which includes different activities that coordinate and work together in manufacturing operations.

Some of the important processes in MES systems are

Quality assurance: It is an activity that documents the quality process, policies, standard operating procedures for the workforce to ensure stakeholders follow the quality process while purchasing, designing, packaging, etc.

Time scheduling: It is an activity to create specific, measurable, achievable, realistic and feasible schedules for each production activity, which is generally done by software program that considers the amount of work, machines and resources needed.

Operations control: It is an activity to track, trace and record production operations which generally involves allocating resources, real time production adjustments, demand forecasting, seamless change orders, procurement, collecting data, etc.

Quality control: It is an activity to ensure acceptable quality levels (AQL) with statistical process control (SPC), nonconformance and event management, corrective and preventative actions (CAPAs), root cause analysis, incoming quality, sampling and testing.

Compliance management: It is an activity to ensure materials are handled according to standard operating procedures to satisfy regulatory and customer compliance. Detailed automatic records with signatures, streamline regulatory reporting and audits.

Equipment maintenance: It is an activity to track equipment usage, preventive and corrective actions (for keeping equipment up and running), materials and spares management, etc.

In this chapter we have understood what is MES, importance of MES in Industry 4.0 and its important processes.

Artificial Intelligence

A.I. is the broad set of technologies (like machine learning, deep learning, neural networks, etc.) which enables devices and machines to think with logic and make decisions.

AI has many branches/ subdisciplines which allows to make computers/ machines to mimic human intelligence and solve problems in Industry 4.0. Some of the important technologies or branches in AI are:

- Machine learning
- Deep learning platform
- Hardware with integrated AI
- Image recognition and Biometrics

Machines learning

Machine learning is a technology which involves algorithms, development tools, API's, models and many more using supervised learning, unsupervised learning, semi-supervised learning and reinforcement learning.

In supervised learning there will be desired output. *You will give an input and tell the expected output. If the output generated by the AI is wrong, it will iteratively adjust the computing over the data set and produces the expected output.*

In unsupervised learning, it doesn't have defined outputs. *Here in unsupervised learning, AI will make logical classifications of the data. An example is an e-commerce portal which will create its own classification of the input data and will tell you which kind of users are most likely to buy different products.*

In semi-supervised learning there are few desired outputs. *An example of semi-supervised learning is the classification of webpages. Since the number of webpages are numerous, you can write a Python/Java crawler, and it can collect and classify the pages in a few hours.*

Deep learning platform

Deep learning is a method in AI which uses the neural network with several layers of nodes between input and output (this series of layers between input & output feature identification and processing in a series of stages). With the help of machine learning algorithms, relationships are identified between complicated data sets (solving problems in areas like computer vision, speech recognition, and natural language processing, etc.).

Deep learning working model

A deep learning process consists of two main phases: analyzing and inferring.

Analyzing is a process of labeling large amounts of data and determining their characteristics, later it compares these characteristics and memorizes them to make correct conclusions when it faces similar data next time.

Inferring is a process of making conclusions and label new unexposed data using their previous knowledge.

Hardware with integrated AI

It is a method in AI where the devices are specifically structured to execute AI oriented tasks.

Examples of hardware integration with AI are:

- Robotic processors integrated with AI
- Mobile phone processors integrated with AI

Artificial Intelligence lifecycle phases

Artificial intelligence life-cycle can be defined in 3 stages as data analysis, discovery and deployment.

Data analysis

Data is the basis for AI in Industry 4.0 as data comes from a multitude of objects like sensors, network data, IoT, cobots, social media, digital customer interactions, etc.

Data Processing

Data processing is the stage where exploring, understanding, modelling, defining algorithms, etc. happens.

Decision making

Decision making is where we actually get value from our analysis, accordingly processing and decisions are made in form of actions.

So now we have understood what is AI, its lifecycle phases and with this, we complete the holistic information on AI.

Blockchain

Blockchain is a tamper-proof technology which stores identical blocks of information across the network, ensuring that a blockchain will not be controlled by any single entity nor does it have a single point of failure (with respect to hacking, unavailability, etc.).

It is a tamper-proof, shared digital ledger that records transactions in a public or private peer-to-peer network. Here each transaction has a unique digital signature which is acknowledged and authorized by the entire network, protecting them from deletion, tampering, and revision.

Blockchain in Industry 4.0 will play a very important role:

- In identifying and tracking the IoT devices.
- Ensuring sensors time-stamp is not manipulated.
- Providing a platform to enable 3D printing through smart contracts.
- Tracking containers during shipping process in supply chain management.
- Recording all important product information throughout the supply chain.
- In cyber-physical systems which can safely and autonomously place an order for their replacement parts.
- In preventing cybersecurity issues (as interconnected devices are more vulnerable to security issues).

Applying the idea of a Blockchain method for Industry 4.0 and its infrastructure creates a data stream with some very attractive properties:

- Consensus across the entire network, on what is true at any given point of time with accuracy and completeness.
- Trust is gained by having an immutable audit trail making it almost impossible to change anything that was written to the blockchain.
- Encryption of data keeps the blockchain visible to only authorized nodes on the network.
- Allows numerous smart machines, devices and things to perform transparent and frictionless financial transactions, without human intervention.
- Product blockchains are very good at telling where a machine is in its lifecycle and who has ownership of it at any given point in time. By using pointers and hashes, blockchains can also tell us which equipment was used at each stage in the life cycle.
- Using smart contracts, defining how the machines should be handled under certain conditions.

In this chapter we have studied what is blockchain and its importance in Industry 4.0 and with this, we will end this chapter.

Internet of things

Anything or any object which has a unique identity like an IP address (to connect, communicate in a social environment with the help of short-range mobile transceivers or any other technologies) enabling new forms of communication between people and things.

IoT in Industry 4.0 plays a very significant role, since the machines, parts and components are connected; now these things (machines, parts, etc.) can tell us a lot of information. For example, IoT in supply chain management would tell what the most appealing products are, what components are being sold mostly, what components have maximum repairs, how they are being used, when they need repair, etc. These things or objects give information to manufacturers which can help at every stage of the production process (from supply, development, assembling, packaging, sales, etc.) to give better insights of the products and customer requirements.

An example from Retail industry, when your retailers' information systems are connected to cloud using big data and analytics you can understand what products are sold widely, which components are sold widely, which products have maximum repairs. When your retail stores have sensors with CCTV cameras connected with the help of IoT, you can get transparent information of your most appealing products accordingly you can do the necessary changes or bring the necessary innovation in your production process.

With the help of IoT, your operational teams and supervisors can get the information about the replenishment in their inventories.

With the help of IoT enabled robots, manufacturing industry can gather information about humidity levels, temperature, vibration and acoustic levels, which feeds back information about real-time machine effectiveness. If any robot gets crashed, the HMI can show us instantly where the robot is located, and what might have gone wrong and remotely switch off power to the robot.

When car manufacturing companies enable IoT, these cars will have the ability to notify the users when it needs an oil change, they can make an appointment on the user's calendar with the dealer for a servicing.

When industries enable IoT on home appliances, the equipment can notify the users on the user-defined conditions and accordingly inform the users.

IoT systems working model

IoT systems work through sensors, local gateways, IoT service, physical devices, cloud storage and big data analytics.

Sensors: Sensors are the small equipment's that can be positioned on the machines, parts, components, etc. Sensors would collect the data aggregate it and send it through a local gateway.

Physical devices: Physical devices are identifiable separable objects of the physical environment like machines, electrical equipment's, electronical equipment's, etc. IoT software deployed on these physical devices provide the capabilities such as monitoring, sensing or actuating.

Note: Sensors, agents, firmware are placed on these physical machines.

Local gateways: Data collected from local sensors is fed into gateways which can be used for real-time monitoring/ analytics.

IoT service: Software components with well-defined and standardized interfaces, enables to access other heterogeneous components with native interfaces.

Cloud storage: Data collected and sent through local sensors and gateways from a variety of machines is stored in the cloud.

Big data analytics: Data stored in the cloud uses big data and analytics processes; based on the rules and thresholds set on these devices, it would enable users or supervisors to act proactively.

In this chapter we have realized what is IoT, its components, working model, examples and with these topics, we complete this chapter on IoT.

Industrial networks and networking protocols

We are very familiar with word 'Network' which is a group of connected computers to share data which was called WWW (World wide web) using HTTP.

Now smart industries will use industrial ethernet to connect industrial infrastructure like SCADA, Routers, and many more components by PLC's.

Industrial networks play a very important role in working the smart factory. As there would be numerous end nodes that would be aggregated for control, monitoring & working 24*7, the industrial network needs robust connectivity and communication. Hence industrial networks have four layers which can be defined as:

1. Field layer
2. Control layer
3. Supervisory layer
4. Execution and planning layer

Field Layer: This layer will accommodate sensors and actuator nodes which can carry data volume in bits (i.e. in 1-100m) with no network topology constraints. These sensors/actuator nodes are typically collecting environmental or process control information and relaying them back to controllers or for monitoring.

Control layer: This layer will constitute Programmable Logic Controllers (PLC's) or Distributed Control Systems (DCS) to provide specific control in the plant. They can execute various types of machines or process instructions through digital or analog modules such as logic, counting, sequencing, timing, etc. Control layer aggregates the sensor/actuator data and transmits the same to larger scale plant control systems.

Supervisory Layer: This layer involves synchronization of PLC's, plant control systems with human-machine interfaces to control the factory operations. In this layer, data volume to transmit would be in Kilobytes and distance can be between 2-40 km.

Systems in this layer are controlled by human operators, through continuous remote monitoring to provide a comprehensive view of the entire industrial operations.

Execution and Planning Layer: This layer is involved in data exchange, computer security, and standards between software packages.

Enterprise and planning related functions such as demand planning, procurement, forecast management and inventory are performed at this layer.

Industrial network protocols are the rules that define the communication between industrial infrastructure and network devices. Some popular industrial protocols that will be used in smart industries are:

1. Fieldbus
2. Industrial Ethernet
3. Wireless

Fieldbus is a network protocol used for two devices communication with each other over serial connections (like RS 232). This protocol can support daisy-chain, star, ring, branch, and tree type network topologies.

Industrial ethernet involves network protocols like Modbus TCP/IP, Ethernet Cat, Ethernet IP, Profinet, etc. they support rugged connectors and extended temperature switches to withstand rugged industrial settings.

Wireless protocols involves network protocols like ZigBee, WirelessHART, 6LowPAN, 802.15.4, etc. which brings many advantages like saving network operational costs, ease of installation, ability to drive more monitoring and control points.

In this chapter we have understood the industrial networks and its protocols and with this, we complete the holistic information on industrial networks.

Middleware technology

Middleware technology in Industry 4.0 is used to connect and manage all the heterogeneous components in industrial infrastructure.
Middleware technology stores, analyzes, and processes huge amounts of data that comes from the networks such as 3G, Bluetooth, etc.

Middleware technology acts as a bridge between objects and applications; it hides the details of the hardware (rather than interacting with the baseline hardware) providing a solution for real time peer to peer communication on standard Ethernet which enhances interoperability of smart objects and makes it easy to make the industry smart.
Middleware provides many advantages like facilitating remote connectivity and remote control, it integrates well with new trends and technologies (like IpV6, Big Data, SOA, Semantics etc.) supporting auto-discovery of devices, devices configuration to network and removal from the network.

There is a long list of middleware technologies (COM/DCOM, CORBA, JAVA RMI, TIBCO, MQTT, AMQP, XMPP, etc.) hence I will focus on the types of middleware based on the functionality and design:

Semantics oriented: This middleware technology type can communicate using different formats of data and devices. It focuses on the interoperation of different types of devices.

Event based: This middleware technology runs based on the device interaction through events (generated by the devices). Events will be generated based on some conditions and accordingly the recipient is notified.

Message oriented: This middleware technology type supports the receiving and sending of messages over distributed applications. It enables applications to be disbursed over various platforms and network protocols easily.

RPC Oriented: This middleware technology type calls procedures on remote systems and is used to perform synchronous or asynchronous interactions between systems.

Service oriented: This middleware technology type is based on service-oriented architecture.

Examples of some Middleware technologies are FiWare, OpenIoT, Mulesoft, WSO2, etc.

Note: There are also some open source middleware technologies which are flexible, but it may have limited support for smart factory devices. As the ultimate goal of middleware in Industry 4.0 is to interoperate with all kinds of communication protocols and devices, hence a proprietary solution is better.

With this, we complete the holistic information in Middleware technology.

RFID technology

RFID has been used in inventory management and asset management for quite some time. RFID in Industry 4.0 provides more efficient manufacturing products at much lower costs, lesser time, with a high degree of automation and with more efficiency.

RFID (Radio Frequency Identifier) technology is an auto-identity data collection system, using RF waves for identifying, tracking and doing management of material and objects flow.

RFID working process

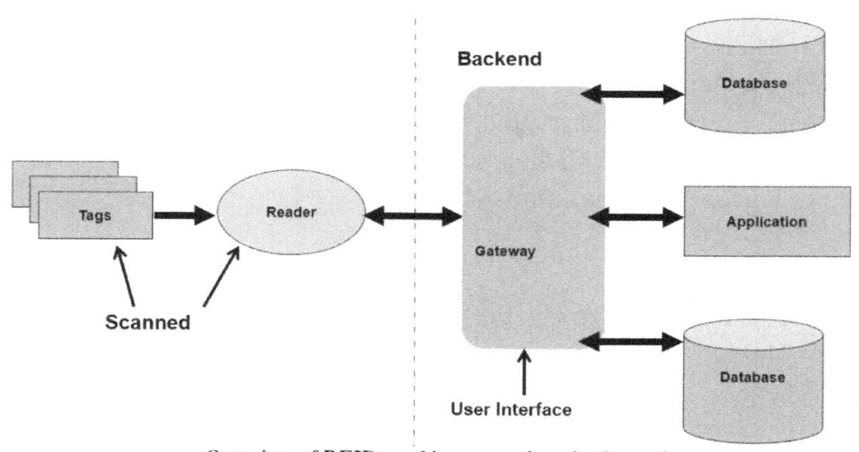

Overview of RFID working operations in three phases

RFID working process can be explained by the below mentioned procedural information.

1. Firstly, items are scanned by reader
2. Second, transmitted data which comes through antenna (RF-wave) is recognized by RFID-based system in the back-end.

(It also acts as a middleware communication gateway among items, reader and system database).

3. Finally, it filters and stores data in RFID databases for checking the data fault and relevant operation.

RFID system components

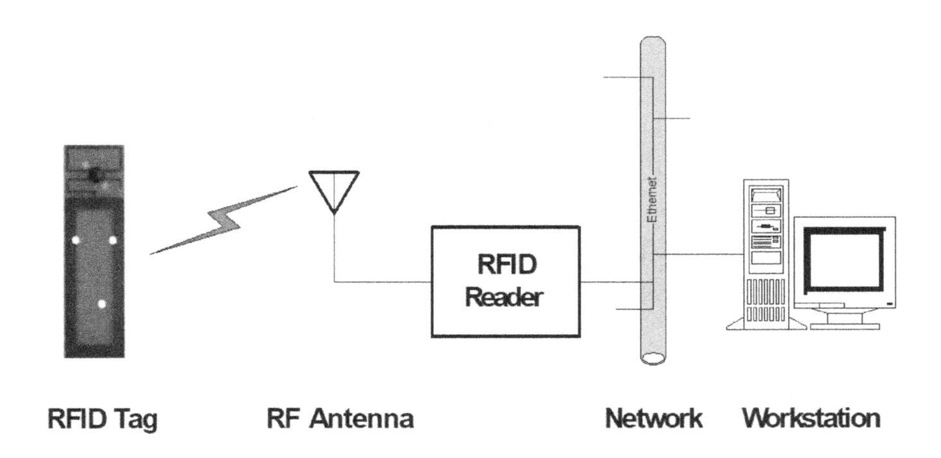

RFID in Industry 4.0

RFID technology in Industry 4.0 helps to:

- Speed up tracking of products.
- Identify their spare parts, making sure that the correct replacement is adjusted.
- Communicate to machines regarding what is needed.

  (For example: After performing a specific task in the manufacturing process, the machine can update product status to the tag, and then send the product to its next stage.)

- Access the data at any point for real-time updates and analysis. The RFID tag's on-chip security mechanisms can serve to prove authenticity, quality, and safety, at any time.
- Verify and access details about the product, its contents, and its use.

RFID types

RFID types available for smart factories are of two types, long range RFID and vicinity RFID.

Long range RFID operates in the UHF (Ultra High Frequency) band, has a read/write range of up to 15 meters, and offers the fastest recognition rates.

Vicinity RFID works in the HF (High Frequency) band, at 13.56 MHz, up to 1.2 meter, and has a lower tag recognition compared to long-range RFID.

RFID tags come in a variety of formats to suit the specific needs of a wide range of industries which can be developed specifically for different industries. So now we have understood what is RFID, working process, components, types and with this, we complete the overview on RFID.

RTLS technology

Real-time location systems (RTLS) is a technology that will be very useful in smart factories which identifies, tracks and locates people or things in real time. It can collect information in either passive or active mode to track the location information. RTLS technology uses wireless technologies (like WiFi, GPS, Infrared, Bluetooth, and active and passive RFID systems) to establish the communication between tags and readers.

RTLS devices typically include location sensors that are attached to various assets (like machines, a staff member or a piece of equipment). Utilizing a unique ID, the system can locate the tags and give you real-time information about its positioning within the facility.

RTLS in Industry 4.0

- Locates and monitors any mobile asset or equipment.
- Gets usage history of any machine (for any time period, department, line).
- Gets real-time location-based view of asset inventory.
- Monitors asset maintenance status and condition.

RTLS working procedure

RTLS working procedure can be described through the below mentioned points:

1) Tags or badges to transmit radio frequency
2) Connects regularly/ periodically to access points

3) Communication is sent to RTLS software.

RTLS components

Tags and badges

Tags and badges provide location information. Tags are attached to assets of an organization that needs to be tracked, badges are worn by people who should be tracked. There are different types of tags available for RTLS, some of the common and prominent are UWB (Ultra-Wide Band) tags, WiFi location tags, Infrared location tags, Passive location tags, and Active RFID location tags.

Access Point / Reader

A wireless access point (AP) is a device which enables wireless devices to connect to a wired network using Wi-Fi or related standards.

System controlling software

This software is installed on server (on-site) where the signal measurements are taken from radio communications and delivered to the RTLS controller. Algorithms built in the software calculates the accurate location of tags and badges and informs the exact location of the assets or people.

RTLS types

Smart factories use two types of RTLS, which can be mentioned as Precision based RTLS & Proximity based RTLS.

Precision based RTLS

Precision-based RTLS allows tracking of assets to an exact location through either ultra-wide band or Wi-Fi-based technologies. Precision RTLS is expensive and requires a complex infrastructure to drive its accuracy.

Proximity-based RTLS

Proximity-based RTLS allows tracking of assets to a proximity within 100 square feet, in-spite of exact locations. These systems are not expensive as precision based RTLS, as it requires simple infrastructure.

So now you understand the importance of RTLS, working procedure, components and types, with this we complete the holistic information on RTLS.

HMI technology

HMI (Human Machine Interface) is a technology that will enable the humans to interact with a machine using a device or software which could be a touch panel on a machine or on a smart phone or any smart wearables.

HMI devices enables visualization and control of applications, by using resources such as I/O, SoftPlc CoDeSys or Ethercat, embedded operating systems, etc., allowing users to communicate with any industrial production systems.

HMI helps users to communicate with machineries and production plants by interpreting complex sets of data into accessible information. With the use of HMI, industrial staff will have all the necessary tools to control the machinery and production process.

In the traditional days, we could think of HMI as a push button on a machine, or a computer with a keyboard. HMI can be considered as a subset of UI (User Interface) which can include input technologies like voice recognition, retina recognition, gesture recognition, etc.

HMI terminals are significantly used in the manufacturing industry for automation to improve the overall plant management by improving communication between user & machine, providing warning & alarms, and cost reduction.

Benefits of HMI in Industry 4.0

- Provides easy, understandable view and access into the industrial systems to control and maintain the machines.
- Provides automation in machines and industrial infrastructure improving communication between users & machines (for example: providing warning & alarms).

- Monitors production and responds to changing production demands.

Examples of HMI in Industry 4:

- Augmented reality glasses, smart watches or wearables which help workers perform their activities faster and more accurately.
- Touchscreens and speech interactions with machines.

HMI in Industry 4.0 can be implemented where human intervention with a machine or automated device is needed. This technology uses sensors, software, wireless systems, enterprise software, machine-to-machine learning or other technologies to gather and analyze data and trigger events to machines.

The three new trends in HMI (Human-Machine Interfaces) technology with respect to Industry 4.0 are:

- Symbiotic HMI allowing dynamic assignment and sharing of tasks and responsibilities between human and machine,
- Social HMI supporting human communication and collaboration between team members for efficient decision-making and coordination,
- Emotional HMI involves in creating positive user experiences as a source for increased human engagement, creativity and flexibility.

Types of HMI

There are different types of HMIs available today, but some of the prominent are pushbutton, data handler, and the overseer.

The pushbutton HMI has streamlined manufacturing processes, which can centralize all the functions of each button into one location. The pushbutton takes the place of LED's, On/ Off buttons, switches that performs a control function.

The data handler will be apt for applications which requires continuous feedback from the systems. In data handler, you need to have big HMI screens for visual representations and production summaries. The Data Handler is used for applications that require constant feedback and monitoring.

The overseer will be useful when an application involves SCADA or MES with several Ethernet ports. These are centralized systems that monitor and control entire industrial operations.

So now you know what is HMI, what is its role in Industry 4.0, trends and types in HMI and with this overview, we complete the chapter on HMI.

In-depth view of Industry 4.0

In-depth view into Industry 4.0

In-depth view into Industry 4.0 covers various topics like stages in setting up I4, foundational pillars for I4, operations, software systems needed, implementation guidelines, predictive and prescriptive maintenance, process maturity framework, prime processes needed in I4, assets lifecycle, I4 site selection factors, green Industry 4.0, quality & auditing in I4, etc.

Pillars for Industry 4.0

Industry 4.0 primarily relies on 4 pillars, which can be mentioned as people, products, processes, protocols & standards.

People:

People in Industry 4.0 perform processes and procedures associated with industrial operations and management. Here, people aspect just doesn't cover the employees, it would involve vendors, manufacturers, suppliers and partners.

Management in industries should understand some important factors with respect to people aspects like:

- Are the employees ready to accept the cultural shift and ready for training, in new responsibilities with new technologies?
- Do they have the skills required to use the new procedures and tools?
- Are there any skill gaps, requiring training?
- Have the customers been identified, along with the customer owner of the service?
- Do the customers understand Industry 4.0 and buy into it?

Smart industry is not just about robotics, cloud, big data; the most important is about providing smart and intelligent workplace for employees. This workplace needs to have improved ergonomics, reducing laborious efforts and increasing the level of individualization.

Smart industries would have many hardware and software solutions that can automatically identify workers, provide them instructions that are oriented to their knowledge, and provide them with smart assembly tools and devices that are intuitive to operate.

Process:

Processes lays the most important factor for Industry 4.0 operations, as all the industrial operations will run based on the processes defined.

To define processes, management should understand:

- What are the measurable and repeatable activities in industrial operations?
- How well do we understand them?
- Are we continuously re-evaluating them for efficiency and effectiveness of operations?
- What capabilities must support them?
- What activities do we have that are not easily measurable, yet still apparently critical?

Some of the prime and generic processes necessary for industries are procurement management, financial management, risk management, demand management, project management, information security management, catalog management, business continuity management, inventory management, disposal management, safety management, outage management, change management, access management, and knowledge management.

Products:

Products refer to the technology, machinery, and software used in industry. Products are further classified into hardware and software.

Hardware used in Industry 4.0 would be:

- Sensors which will be used in manufacturing plants and surrounding environments (in massive numbers to provide production data).
- Standard communication networks that enable plant-wide data collection and communication across the factory floors and supply chain.
- Robots which play a major role in making manufacturing processes more productive and less labor intensive. (Robots are typically responsible for sensing, motor driving, and movement functions that are like routine tasks.)
- Remote I/O devices that communicate with sensors, actuators, networks, etc.
- Cameras to capture images for pattern checking, barcode scanning, defect inspection, measurement, and also inspecting employees' activities.
- HMI panels enabling operators to interact with the production equipment.
- CPS gateways communicating downstream to manufacturing modules over various fieldbus protocols and upstream to an on-premise SCADA system or to the cloud (or data center) via implemented networks.
- IoT gateways providing end-to-end connectivity for monitoring and maintaining manufacturing assets. IoT gateways extract information from field data and transfer the information to the cloud for analytical, archival, or other purposes.

Software systems refer to the software and technology that will be used in machinery to communicate with the users to make better decisions. Software that is generally used in Industry 4.0 would be:

- Distributed control systems to flexibly connect distributed I/O's, sensors, and drives. Distributed intelligence is a basic requirement for modular machines and flexible facilities that adjust themselves to changing market and manufacturing conditions.
- Plug and play software enables machines in a network on a regular and ad-hoc basis, this software simplifies multiple smart manufacturing machine procedures like installation, commissioning, integration and de-commissioning. (Also, preventive maintenance of all components and machines)
- With the help of simulation software, all components and objects are represented as virtual objects across the entire value creation process. These virtual elements are closely linked to their physical counterparts and provide in-context information for continuous process improvement in real-time.
- Production software enables long- and short-term planning; predictive control; optimization; environmental, health and safety management; and other intelligent services about manufacturing operations.
- ERP & MES software to make better product decisions and improve development by ensuring stakeholders have the latest and most accurate product information.
- Remote monitoring software to solve common problems in the factory and empowering staff to monitor assets with real time alerts and reduce costly downtime.

Protocols and standards:

Protocols & standards provide the ability to seamlessly exchange electronic product, process and project data between collaborating groups or companies and across the business systems.

When different machines, sensors, systems use different protocols and technologies to transfer the data, it becomes difficult to get them to communicate with each other. Hence having one single industry protocol facilitating communication between devices and software enables machines, sensors, gateways, etc. to communicate better.

Open Standards that extend across manufacturers are platform-independent which forms the basis for horizontal and vertical integration and thus for the seamless exchange of information in value-creation networks.

The above mentioned four pillars (people, processes, products & protocols) are immensely involved in the life-cycle of smart industry which can be represented in 4 activities as:

Data collection → Data transmission → Data storage → Data reporting

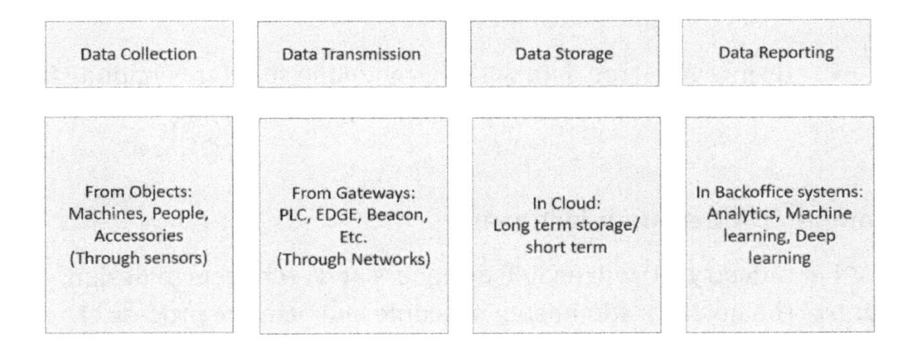

Software systems in smart industries

Software systems plays a very important role in Industry 4.0, to make the machines work smarter and intelligent. Some of the key software that will be used in smart industries would be:

Computer Aided Engineering

CAE in Industry 4.0 will enable designers, engineers and analysts to simulate products, provide cost estimation and design optimization tools. With the help of CAE:

- Users can make designs faster and accurate.
- Users can identify issues in minutes,
- Enables users to implement changes earlier in the product life cycle, which ultimately saves costs.
- Users can merge data with 3D modeling software enabling to generate virtual prototypes before manufacturing.

Computer Aided Manufacturing

CAM in Industry 4.0 will enable designers and architects to design, control (on quality), administer, schedule and plan the process of manufacturing. With the help of CAM:

- Designers can understand the information of production equipment and facilitate the production of efficient machining.
- Maneuvering the parts to select features is very easy.
- Designers can plan data, obtain services of production, give a better view of the finished product.

Statistical process control software

Statistical process control software in Industry 4.0 simplifies the manufacturing process to ensure that you get the end-to-end visibility with accuracy.

Smart factories will have numerous machines, and manually gathering data from these machines involves lot of time and is a tedious job. With real-time statistical process control (SPC) software, industries can automate the data collection process.

Statistical process control software increases user productivity, optimizes the usage of machines, prevents defects, reduces scrap, and reduces rework.

SPC software helps you identify and eliminate variations in manufacturing process, ensuring:

- stability to consistently produce the same results.
- capability to meet the design specifications.

Bigdata analytics

Bigdata analytics in Industry 4.0 will enable the staff to track the health and configuration of all smart devices and machines, which can:

- identify asset maintenance needs and even predict failures before they happen.
- analyze the data and identify patterns of behavior.

This will enable:

- Awareness of the health of the machines.
- To reduce machine downtime and increase uptime.

It will also help the industrial operations, by collecting performance data like mean time to repair, mean time between failures, time taken for assembling a machine, etc.

Manufacturing Execution System

MES provides the overall control and management of the factory floor and provides updated information to the ERP.

MES software generally has functions like operations scheduling, data collection, product tracking, document control, resource allocation, etc.

MES software enables to:

- Bridge the gaps between ERP and the shop floor,
- Eliminate the flow of paper documentation,
- Optimally use manufacturing resources,
- Provide better view of the whole manufacturing process, synchronize materials and operations,
- Interlock traceability and genealogy covering materials (components and finished goods).

Infrastructure Visualization systems

Infrastructure Visualization systems provides a clear and transparent view describing what is happening, why it is happening and when did it happen on the shop floor.

For example, when a machine is not working, it gives information: on the reason of the outage, time to resolve the outage, who is responsible for the fix.

Tracking solutions

Tracking solutions are used to record the product/ products manufacturing process with unique identification and labeling system.

These systems guarantee the quality and possibility of reference to a specific batch of products (that uses batch tracking and production registration to monitor production efficiency, utilization of production buffers, determination of actual course of operations and their duration).

Sampling solutions

Sampling solutions measures data from the technological processes and ensures correctness, appropriate frequency of checks and introduction of repetitive procedures recorded in the computer systems.

Approach for implementing smart industry

Redefining a factory into a smart factory is not an overnight task, it involves a streamlined approach and various procedural steps, here are my few cents to make your factory into a smart factory:

1. Firstly, make your C level officers understand and convince the impact of I4.0. They have to be aware of the benefits that Industry 4.0 can offer and understand the use cases, value of data and the importance of IoT, cloud, big data, cyber physical systems, cyber security, etc.
2. Identify, prioritize and pilot the most important operations that can help and increase productivity, reduce time and operating costs.
3. For each use-case, proof of concepts should be built to explore suitability for the organization using simulations and modeling software.
4. Understand the existing inputs and suppliers for manufacturing operations.
5. Understand the existing processes, automation, outputs delivered by the manufacturing operations.
6. Understand what the most important machines are (HVAC, power, production machines, etc.) and the machines which perform routine activities, which requires monitoring and controlling.
7. Implement equipment connectivity: devices, machines, production lines and factories so that the visualization can be fulfilled. Accordingly implement cobots, sensors or custom-built sensors for all the important machines and its components.

Note: You can consult companies who would install hardware (sensors) and software to your infrastructure (manufacturing operations management, logistics, supply chain, disposal, etc.) which will give access to the data generated by the industrial processes.

8. Introduce data collection and integration, by implementing IoT platform and gateways to gather, process, analyze the information and to communicate with devices.
9. Understand what data is generated, which machines, which activities, and how the data is generated.
10. Implement big data analytics solutions for all the factory operational data.
11. As all your data will be stored in the cloud, here it will be translated into uniform values and processed into understandable information.
12. Now this whole meaningful data can be viewed, analyzed to make better decisions, to reduce costs, to increase the quality of the products, to reduce the time taken for production, to reduce the power consumption, to give better experience to the customers, etc. making the factory into a smart factory.

    The most important in the above-mentioned implementation steps is to, review and revise at regular intervals. As this would be the best way to track how these initiatives are progressing and how they are impacting your industry.

Operations in smart factory

Operations in smart industry is generally handled intelligently and efficiently by leveraging the capabilities of advanced materials, additive manufacturing, robotics, distributed networks, advanced sensors, remote controlled operations, smart machines, artificial intelligence, augmented reality, digital infrastructure, cloud computing and many more.

Here is an example how the customer orders are placed, handled and managed in smart factory:

- Once the customer places an order in a smart factory, the order information is communicated to different business processes and information systems like Enterprise Resource Planning or Manufacturing Execution System.
- With the help of simulation software or augmented reality, product/ service models are designed.
- After the models are designed, this information is sent to internal stakeholders (like supervisors, management) and external stakeholders (like suppliers).
- Once the information is received by the internal stakeholders, the operational staff does the necessary arrangements with respect to machines, equipment, staff, etc. At the same time, external stakeholders also receive information listing the materials needed for production. Together, manufacturers and suppliers form a value enabled network.
- Once the model designs are sent to all stakeholders including customer, the production process starts with the parts, components and machine manufacturing. All the needed components get an ID-tag which enables these components to guide themselves through the factory. This ensures that all

products will be in the right place, at the right time, for efficient manufacturing.

- When the equipments and parts qualifies the simulation validations & human validations, its ID-tag is scanned and that data is transferred to the cloud server.
- This platform gathers all the customer's order specifications, simulations, models and displays what will be produced. This platform collects, processes and visualizes all relevant production line data in real-time – connecting man, machine and production process in a new kind of automation.
- Smart industry uses autonomous robots to connect man and machines working together in routine tasks like assembling the parts.
- Smart industry uses assembling tools that have smart technology to fix the parts and machines better.
- Imagine if there are any modifications in the quantity or specifications of the machines or parts, with the help of smart technology all activities are handled in within the same delivery time. The configuration change is made and confirmed to the customer, and orders for the new parts are sent to suppliers in the value creation network.

This is the holistic and bird's eye approach on the operations of smart industry.

Predictive and Prescriptive maintenance in Industry 4.0

Predictive and prescriptive maintenance will change the dynamics of industrial manufacturing and maintenance operations.

In the earlier phases of industrial manufacturing, (first industrial revolution to fourth) maintenance procedures had a great metamorphosis starting from reactive → preventive → condition based → and now predictive and prescriptive.

Hence let's look at the overview of predictive and prescriptive maintenance in Industry 4.0.

PREDICTIVE MAINTENANCE

Maintenance tasks in industries requires meticulous efforts from engineers and supervisors to perform the repairs, replacements, installations, etc. But with the inception of Industry 4.0 which uses big data analytics, to collect data and help the staff understand the conditions of the infrastructure.

Predictive maintenance is one such benefit in Big Data analytics where patterns or messages or events triggering from Industry 4.0 infrastructure is discovered, analyzed and appropriate decisions are made for maintenance.

Predictive maintenance looks at past data and tries to mine and discover models. Data is available from sensors that are generated from machines which track the operating conditions of the equipment usage, issues that might occur, inform about the servicing, and also predict when the outages might occur.

The most important thing of predictive maintenance is to continuously improve the overall equipment/device effectiveness.

Example of predictive maintenance:

Alarms are generated, logged and delivered to the mechanics in smart factory through mobile devices and these alerts are stored in in enterprise asset management tool. Accordingly, the management can see exactly how a machine, or a part is functioning. Eventually this would lead to predictable maintenance and a longer lifespan of the machines.

PRESCRIPTIVE MAINTENANCE

Prescriptive maintenance is another advancement that will be used in Industry 4.0, which defines what has to be done, when it has to be done and who has to do it (with the help of Artificial Intelligence (AI) and machine learning) providing use cases for operations in Industry 4.0. With prescriptive maintenance, operators and devices will be in collaborative and proactive approach providing a seamless work culture with end-to-end visibility, effective and efficient industrial operations.

Some capabilities of digital prescriptive maintenance include:

- Assembling/ dis-assembling a machine.
- Gathering data from the device or ping it for specific measurements and analysis.
- Monitoring a defective device from the supplier to the customer via GPS.
- Dispatching service technicians based on the repair works.

Key business drivers for prescriptive and predictive maintenance are:

- Automation: Enabling the industrial operations to execute at a quicker pace and make quicker decisions.
- 24*7 Work culture: In 24*7 work culture, all the workforce and machines will work continuously to provide seamless communication between employees.

Essentials for Executing Industry 4.0 Operations

Essentials for execution of Industry 4.0 operations can be defined by the below mentioned activities:

1. Definition of governance structure, roles, and responsibilities
2. Definition of escalation procedures
3. Definition of operations procedures
4. Definition of relationship procedures
5. Definition of knowledge management procedures
6. Definition of auditing procedures

Definition of Governance Structure, Roles, and Responsibilities

Definition of governance structure, roles and, responsibilities ensures that industrial operations are adhered to objectives and goals providing clear segregation of who is supposed to do what.

For example:

- Governance ensures that only privileged personnel can access servers and critical business machinery.
- Governance ensures correct sensors are configured on the machinery and networks.
- Governance on cybersecurity is essential to preserve market positions.

Definition of Escalation Procedures

Definition of escalation procedures will ensure and will enable to get the issues noticed by the management at the right time and by the right people.

For example: Here is a matrix defining the issues and the responsible person who should handle the issue:

Issue	Level
Industrial operational and availability issues	Level 0 – Operations manager
Process related issues	Level 1 – Process Manager
Human resource availability issues	Level 2 – Project Manager
Risk related issues	Level 2 – Risk Manager
Project plan related issues	Level 2 – Project Manager

Definition of Operational Procedures

Definition of operational procedures needs to be there for every department in a smart factory, but primarily for the below mentioned areas:

- Contract management
- Outage management
- Change and configuration management
- License management
- Business continuity management
- Information security management
- Facilities management
- Quality management

Definition of Relationship Procedures

Relationship procedures should primarily focus on:

- Regular interaction with the customer and stakeholders, by participating in meetings with customer (with respect to project reviews, analysis, forecasting, and discussions to get better understanding of the needs).
- Understanding customer's technical, business environment, and their needs.
- Conducting weekly/monthly reviews.

Definition of Knowledge Management Procedures

Knowledge management procedures should focus on:

- Documenting and profiling the work done and organizing lessons learnt sessions.
- Establishment of a database for knowledge articles.
- Organizing the knowledge articles with knowledge article registration number, keywords, error, version, created by, etc.

Definition of Auditing Procedures

Auditing procedures should address the aspects of:

- Personnel procedures and responsibilities, address employee termination, cross-functional, and systems training.
- Program change controls that are adequate to ensure that changes are tested and approved before being moved into production status.

- Backup procedures that are adequate to minimize business interruption and protect against loss of data in the event of a disaster.
- Physical security controls that are adequate to prevent unauthorized access to computer center areas.
- Environmental controls that are adequate to minimize hardware/software losses from fire or flood.

Stages for setting up Industry 4.0

Setting up Industry 4.0 is a huge task which involves many activities, hence I have classified these numerous activities into 4 stages as smart planning, designing, manufacturing and logistics.

Smart planning:

Smart Planning allows you to plan and schedule your orders subject to most diverse dependencies, creating transparency along your entire planning and ensuring a quick and smooth order flow.

Smart planning does the production planning and scheduling focusing on innovative multi-resource planning based on limited finite capacity (machines, personnel, fixtures, and materials). It has thorough control on detailed order scheduling and also provides visualization of results on a graphical control station. It also improves on-time deliveries and reduces total cost ownership (TCO).

Smart designing:

Smart designing is an important step in smart industry automation which relies on CAD/CAM/ CAE systems, statistical process control software, etc. providing parts information and simulating the equipments designs.

With the help of digital designing, running-times on the machines are reduced up to 90%, machine damages are avoided, and your engineering staff can spend more time on production.

Smart Manufacturing:

Smart manufacturing is the heart of Industry 4.0, which uses robotics, sensors, big data analytics, etc. to make manufacturing operations more intelligently as mentioned below:

- Using the integration of machine vision and motion, robotics and AGVs (Automated Guide Vehicles), manufacturing processes can reduce dependence on manpower, reach lean production and even unmanned production to automate manufacturing.
- Sensors are installed on machines to constantly monitor essential production parts, machines and the production status. Data can be recorded and transmitted in real-time to the cloud for predictive maintenance analysis to increase productivity and reduce maintenance costs.
- Big data analysis can be conducted by using frequency, time, temperature, and other required parameters to provide preventive maintenance recommendations.
- Automatic detection systems will be implemented to improve test efficiency and product quality.
- Real-time monitoring systems will be implemented in manufacturing premises for production efficiency, yield rate, quality and achievement rate of the production.
- MES (Manufacturing execution systems) can be implemented to collect dynamic production process information, optimize production allocation, achieve advanced manufacturing and trace the production information. Furthermore, this production process information can be monitored remotely.
- With the help of sensors, industry health and safety can also be monitored for dust, gas, and other hazardous materials to optimize factory operation and ensure the quality of the factory environment.

- With the help of sensors and monitoring systems, industrial power consumption can be optimized and also CO2 emissions can be monitored.

Smart logistics

Smart logistics will ensure the manufactured products/ services are seamlessly transported using autonomous vehicles, tracks and traces movements of products and services, processes unprecedented amounts of logistical information and detects risks regarding logistics more effectively and intelligently as mentioned below:

- Radio frequency identification (RFID) and Bluetooth technologies will be used in inventory for tracking the movement of products/ services.
- GPS technology will be used to check exact shipment locations, trucks will communicate their position and arrival time to the intelligent warehouse management system, which will choose and prepare a docking slot, optimizing just-in-time and just-in-sequence delivery.
- RFID sensors will be used to track what has been delivered and send the track-and-trace data horizontally across the entire supply chain.

Industry 4.0 Process Maturity Framework

What is a maturity model?

Maturity model is a business tool or methodology used to assess their current or as-is operations and processes and to identify the gaps in operations and processes (with respect to people, processes, and products) for effective operations in overall business.

Maturity model generally represents 5 stages as:

1. Chaotic
2. Initial
3. Repeatable
4. Defined
5. Managed and Optimized

Industry 4.0 Maturity Model

I4PMF (Industry4 Process Maturity Framework) is a framework to assess the maturity of industrial operations in Industry 4.0. It defines a methodology which allows assessing the Industry 4.0 processes and operations through standard best practices from IT Service Management, IT Governance, and Project Management. Maturity is indicated by the award of a particular maturity level like 0,1,2,3,4.

Maturity Level 0 - This generally denotes the operations where:

- all work is done manually
- data storage is unorganized

- many errors occur due to manual work
- firefighting happens after an issue or a discrepancy happens
- some tools exist but no defined process
- cannot effectively identify industrial assets, where the assets are located, or how devices are connected without significant manual intervention.

Maturity Level 1 - This generally denotes the operations where:

- some software systems are used (where some systems communicate with each other, some are isolated, and some are incompatible with other systems)
- reporting systems are used to consolidate information about the operations
- minimal machine to machine integration and communication happens
- digital modeling and simulation is used very minimally
- industrial operations rely on the heroics of some individuals

Maturity Level 2 - This generally denotes:

- defined and proactive operations where all the software systems are integrated
- machine and asset tracking is done using software
- sensors and wearable devices are available in some areas
- some amount of data is available in mobile systems, and minor robotic automation is available
- there is defined processes and respective tools for security, inventory, disposal management, and quality management

Maturity Level 3 - This generally denotes:

- where the robotic automations are available significantly
- data is available on the cloud
- big data analytics are used
- sensors are available in most of the machines and infrastructure
- mobile software compatible with many machines and devices
- here the operations are well managed ensuring the services are delivered with respect to defined SLA's, processes are robust and integrated across the industrial operations

Maturity Level 4 - This generally denotes optimized operations as mentioned below:

- can fulfill business requirements in the requested time, quality and cost.
- all the industrial operations can be viewed through monitoring systems
- data analytics are used throughout the value chain, robots for all routine works
- it can perform predictive and prescriptive maintenance
- sensors and wearable devices are available
- data storage is organized and streamlined
- not many errors occur
- machine to machine integration and communication happens
- digital modeling and simulation is used very minimally
- adheres to digital compliance policies

- Intellectual property (IP) for all digital products and services is protected.
- it will have smart factory status and running operations as per the Green Industry practices considering the environment and efficiency in managing the assets with the focus on cost reduction.

Tidbits to make your factory a smart factory

Today many factory management is aspiring to make their factories as smart factories, so here are some of the parameters which should be there to redefine your factory as a smart factory.

- Digitization of manufacturing life-cycle phases (design, planning, engineering, production, services & recycling), horizontal value chain, vertical value chain and services (e.g. RFID for identification, sensors, IoT connection, smart products etc.).
- Should analyze and track the usage of customer data, product data, and machine generated data.
- Should collaborate (exchange of information) with all stakeholders (like partners, suppliers and clients) for development of products and services.
- Various digital and non-digital sales channels (e.g. store, sales force, web-shop, sales platforms etc.)) should be integrated to sell the products and services to your customers.
- Website, blogs, forums, social media platforms etc. should be integrated into customer interactions for communicating news, receiving feedback, managing claims etc.
- Should capture customer preferences based on location, trends, seasons and send personalized offers.
- Should gather data from unified systems to track, assess and optimize products, sales and customer experience.
- Should work together with partners regarding the tactics of accessing customers (exchange of customer insights, coordination of marketing activities etc.).

- Digitization of vertical value chain (from product development to production) and horizontal value chain (from customer order to production and logistics) should be available.
- Should have real-time view on manufacturing lifecycle operations and should be able to dynamically react on changes in demands.
- Should have an end-to-end IT enabled planning and steering process from sales forecasting, over production to warehouse planning and logistics.
- Should have cloud enabled manufacturing execution system (MES) to control manufacturing process.
- IT systems should have the capability to gather, aggregate and interpret real-time manufacturing, product and client data.
- Should use new business trends like social media, mobility, analytics and cloud computing for enabling better business.
- Should institutionalize collaboration on Industry 4.0 topics with external partners such as suppliers, customers, universities, etc.

Prime processes for Industry 4.0 Management

Smart Procurement Process

Smart Procurement management defines standardized procedures for identifying the necessity of purchasing and replenishing the industrial assets. With the help of big data and analytics and cloud technologies, procurement team will be able provide the information of the manufacturers who are economic in cost, and also providing good quality and utility.

Based on the information provided by the procurement, below mentioned activities will be carried out:

1. Background check on the suppliers
2. Selection of the supplier
3. Negotiate and define the agreements
4. Initiate the business operations with the supplier
5. Maintenance
6. Renewal or termination of the agreements

Smart Inventory management process

Smart inventory management process defines a standardized process and procedures for stocking the assets at the right place (containers, shelves, etc.), based on the assets size, specifications, etc. which were captured at the time of modeling and simulations. With the help of CCTV cameras, sensors, cognitive software, the complete industrial area's space is captured and accordingly the software would suggest the right and empty place for the produced assets/ machines in the inventory without unnecessary motion of assets.

This arrangement of assets in the inventory can be done in many ways, some of the most prominent are:

1. Alphabetically: In this arrangement, assets are arranged in an alphabetical order with respect to the name of the asset.
2. Date based: In this arrangement, assets are grouped based on the date of purchase, delivery dates, etc.

Smart Disposal Process

Smart disposal process defines and manages the standard procedures for informing the EOL (End of Life) date for every machine. With the help of sensors equipped on every machine, EOL data is captured in the information systems and notifications will be sent to the relevant asset owners to dispose the assets.

Also based on the reliability, capacity, and performance of the assets, the cognitive software would be informing the right methodology to dispose the assets as per the organizational, legal, and environmental requirements.

As improper asset disposal methods can cause severe fines, lawsuits, and irreparable brand damage, smart disposal process would inform the asset owners to dispose the assets at the right time and also the right methodology.

To avoid any penalties and to mitigate risks involved in the asset disposal process, companies need to proactively define procedures for retiring and disposing of the industrial assets and this will be possible with the smart industry.

Disposal process is triggered when asset sensors detect the below mentioned disposal criteria like:

a) The assets become inefficient to operations with respect to availability, performance, etc.
b) The assets have negative impact on the day to day operations.
c) The assets become hazardous to use.
d) The assets have met or crossed the expiration date.

Methods for industrial asset disposal can be done based on the asset's functionality, associated financial value, asset utilization, asset average lifetime, and other factors. Generally, industries discard as scrap, after removing the toxic materials considering SHER (Safety, Health, Environment, and Risk) standards.

Smart Risk management process

Smart risk management defines standardized procedures for identifying the assets, analyzing the threats, analyzing and assessing the risks using IoT, cloud, analytics, concepts as per the defined security objectives. Based on the information gathered, risk manager will be implementing the countermeasures.

Activities involved in industrial risk management are:

- Prioritizing risks, defining policies and automating assessment processes for IT and OT/ICS environments.
- Enforcing IT policies and automating compliance (ISO 27001 & 27005) – with built-in automation and workflow to not only identify threats, but also remediate incidents as they occur or anticipate them before they happen.
- Communicating IT and OT risk using information security standards.

Smart Safety management

Smart safety management is a process which defines the procedures and activities for working, using, storing, manufacturing, handling, auditing and evaluating the operations in an industry.

Purpose of safety management is to identify, understand and control process hazard risks, creating systematic business improvements and safety standards.

Security and safety for Industry 4.0 includes protecting people from machinery-related hazards (safety) as well as the protection of production facilities from attacks and faults from the surrounding environment (security). This also includes securing sensitive data as well as the prevention of intentional and unintentional malfunctions.

Industrial Assets Lifecycle

Industrial assets lifecycle would generally involve 11 phases as mentioned below:

Plan

This step involves understanding the requirements from the internal/external customer with respect to functionality, power usage requirements, air conditioning requirements, cables, transactions per second, etc.

After consolidating the requirements, it has to be brainstormed with SME's. Then after technical and financial approval, it moves to the next step requisition.

Requisition

This step is triggered by the purchasing team sending a request for new machine. The inventory team checks if there are any assets available in the inventory with the specified configuration and will decide to purchase a new asset or use any purchased asset in the inventory.

Procurement

This step involves procurement team's work in obtaining the quotation from different suppliers and choosing the best one ensuring value for money.

Acquisition and Stocking

This step involves acquiring the assets from the suppliers, verifying whether if we have received the right assets as per the requirements and stocking it in the inventory.

Discovery

This step involves discovering the configuration of each machines and its components, and the complete machine so that it can be configured and installed correctly.

Qualification

A new equipment should be initially quarantined by assurance checks on three aspects: authentication, scanning to detect if there are any damages, and compliance checks to ensure that it suffices the policies.

Bare Metal Provisioning

This step prepares the machine or equipment for installation and involves a variety of tasks such as configuration, tuning, and loading basic system configurations.

Service Provisioning

This step would assign the machines or equipments to different departments or services with necessary accessories and applications like sensors, software, network/storage access, etc. so that they can provide intended services.

Monitoring

This refers to constant monitoring and maintenance of equipment taking suitable actions. Monitoring data from various equipments like machines, robots and software elements would typically involve filtering, storage and fusion in order to detect and resolve problems, minimize power consumption, and determine security attacks, etc.

Remediation

This refers to activities related to fault detection/diagnosis, security related quarantine, repair, upgrade and replacement. Remediation may be required while the equipment is in use and thus may interfere with the service.

Retirement and Disposal

This refers to activities related to the disposal of retired and obsolete assets as per the organizational, legal, and environmental requirements.

Industrial equipment will be disposed when the equipment meets the below mentioned criteria:

1. The equipment is not required for the organization due to changed business goals and end products.
2. The equipment has become inefficient to operations.
3. The equipment has negative impact on service delivery.
4. The equipment become hazardous to use.
5. The equipment has met or crossed the expiration date.

Industrial equipment disposal can be done based on the asset's functionality, associated financial value, asset utilization, asset average lifetime, and other factors. Some of the well-known methods are:

1. Reassignment to less business-critical operations
2. Selling in a public forum.
3. Discarding as scrap after removing the toxic materials considering SHER (Safety, Health, Environment, and Risk) issues.

Site selection factors for Smart Industries

Site selection is one key success for stable and continuous business operations in smart industries. There are some important and critical success factors to consider before choosing a site for a new industry as mentioned below:

- Electric Power: Power is typically one of the highest on-going costs of operating a facility; understanding the vulnerabilities and reliability of the power infrastructure components is an essential part of site selection. It requires good availability, economic cost, and redundancy for a smart factory to run the operations. Choosing a location with desirable electrical rates is an important factor. Clean power is also a key criterion in the site selection process. Industry owners/operators seek reliable, efficient renewable energy options such as hydro, air, wind and, to a lesser degree, solar. Fuel cells are also emerging as a viable power option.

- Risk of Natural Disasters: An area that is prone to natural hazards may not be a suitable for a large building location. Seismic events, floods, tornadoes, hurricanes, and volcanoes can place the smart factory as well as suppliers of power and other services at risk. Weather events and natural disasters have the potential to interrupt operations and are of great concern when locating data centers.

- Weather: Weather and climate can significantly affect the efficiency of industry cooling, which in turn affects sustaining costs. Temperate and climates are a growing subset of the environmental factor thus driving down PUE (power usage efficiency) - a major consideration in industries today.

- Land & Construction Costs and Quality - The cost of the land is a major criterion. It is better to choose sites with an appropriate cost per square meter (acre) and prefer sites with single owners because multiple owners can slow negotiations and purchasing activities. Also, it needs availability of qualified construction labor, such as framers, concrete workers, electricians, and plumbers, contribute to a successful construction project.

- Quality of Life – Industrial employees will be involved with a mix of labour from highly educated and basic schooling education hence quality-of-life factors are an important consideration in attracting and retaining all the kinds of employees.

- Labor Costs and Availability: Availability of workforce is very essential factor as smart industries typically would need thousands of employees and as factory owner, you need to ensure that the employees have reasonable commuting distance.

- Availability of Telecommunications Infrastructure: Various types of media, such as fiber, copper, and satellite, are used in WAN connectivity for IoT and Cloud technologies. Understanding latency limits and the location of preferred fiber providers and telecommunication points-of-presence increases the overall success of the site selection process.

- Tax and Regulations: The local, regional, and national taxes, regulations, and incentives can affect virtually every consideration for choosing an industry location. Taxes may include sales, personal property and real estate. Regulations can be enforced at the local, regional, and national levels. Regulations that should be considered when choosing an industry site include land zoning, fuel storage and emissions, etc.

- Transportation: Proximity to prime corridors can be viewed as favorable condition as access to major transportation modes, such as airports, rail lines and interstate roadways, creates the potential for easy transportation link.

- Proximity to Water Resources: Water is a key resource for smart industries because drinking water, water for air conditioners, water for machines, etc. is one of the cost factors.

Parameters for industry site selection are:

Parameter	**Availability condition**	**Remarks**
Risk of natural disasters		
Weather		
Land & construction costs		
Labor costs & availability		
Transportation feasibility		
Power Reliability		
Power Quality		
Alternative Energy Options		

Electricity Rate Price		
Power Infrastructure Cost		
Power Infrastructure Timing		
Power Available Capacity		
Percentage of Utility Load		
Utility Financial Stability		
Privatized/Deregulated/Monopoly		
Power Customer Service		
Power State-of-the-Art Equipment		
Future Stability of Power Resource		
Power Line Specifics		
Water Reliability		
Water Infrastructure Cost		
Water Infrastructure Timing		
Sewer Reliability		
Sewer Infrastructure Cost		
Sewer Infrastructure Timing		
Natural Gas Reliability and Capacity		
Natural Gas Rate Price		

Green Industry 4.0

Industries that use energy-efficient machines, computing devices, lighting equipment, electrical equipment targeting reduction in energy demands is called as green industry.

The primary objectives of green industry are:

- To maximize energy efficiency
- Reduce energy costs
- Reduce carbon emissions and environmental impact

Best Practices for Green Industry

- Block cable openings to prevent cold air waste in the hot aisle.
- Removing under-floor cable blockages that impede airflow.
- Turning off machines that are not doing any work.
- Turning off machines room air conditioning units in areas that are overprovisioned for cooling.
- Organizing industrial equipment into a hot aisle and cold aisle configuration.
- Positioning the equipment so that you can control the airflow between the hot and cold aisles and prevent hot air from re-circulating back to the industrial equipment cooling intakes.
- Procuring smart energy machines and equipments, considering humidity, temperature and compliance with green regulations.
- Installation of catalytic converters on backup generators.
- Use of alternative energy technologies such as photovoltaics, heat pumps, evaporative cooling, low-power lamps, and lighting sectors.
- Use of low-emission building materials, carpets, and paints.

- Use of Sea Water/Recycled Water and reuse of heat into office buildings where applicable.
- Cooling the IT equipment within the racks.
- Using natural ways to cool machines.
- Collection of data through monitoring systems to determine the state of your current systems. Determine how much power is consumed by machines and break down the costs.
- Energy usage should be continuously monitored to determine peak and low energy demands.
- The energy savings plan should be documented and rewarded. Also, it should be reviewed regularly with corrective action taken to address failures.
- An inventory of your current systems and tracking their power usage and locations.
- Track your company's business and growth plans - to help forecast future needs.
- Establish goals for reducing your company's carbon footprint - and the timeframe set for achieving those goals.

Auditing Industry 4.0

The objective of audit is to ascertain the effectiveness of existing policies and procedures related to the administration and control of industrial operations. Also, it is to determine the adequacy of controls over the related processes and ensure compliance with relevant governmental regulations.

Key focus areas of the audit are to ensure:

- Industrial policies and procedures are defined, documented, and communicated for all key functions.
- Council systems are secured to prevent unauthorized access to industry premises and machines (including 3rd party access).
- Access to the industrial premises, machines and autonomous robots is monitored and reviewed, and access rights are periodically reviewed.
- Data stored in cloud is secured at all times and appropriate controls are in place to monitor the location of the data.
- Environmental controls are present to protect the machines from fire, electrical and water damage.
- Environmental equipment is routinely maintained in line with manufacturer recommended schedules.
- Backup electricity supplies are in place to ensure systems and services are not affected in the event of a power outage.
- Background checks are performed on all employees who have physical access to very costly machines.
- There are procedures to monitor and review electronic key card access system reports that record the time, dates, and names of employees entering and exiting the industrial premises.

- Industrial operations staff should be trained on emergency evacuation procedures, fire safety, and the use of fire extinguishers.
- Service Continuity Plans should be updated, tested & documented.
- Location of the prime machines is away from risks like fire and explosion hazards, water related accidents, etc.
- There is an updated inventory of equipment, systems, and data residing in the factory.
- There is record of information to rank the systems' criticality and establishing a priority with respect to the business value and financial impact.
- There is a cost analysis conducted with implementing or improving controls.
- Checks on electrical equipment happening on a regular basis for capacity planning
- Energy savings plan is documented and rewarded.
- Energy savings plan is reviewed regularly, and corrective action is taken to address failures.
- Energy usage is continuously monitored to determine peak and low energy demands.

Challenges in Industry 4.0

Smart factories involve very complex IT systems, which in turn faces a great number of challenges as mentioned below:

- Defining business models and new strategy for the smart factory.
- Huge investments are needed to make a traditional industry to Industry 4.0, and this journey is not cheap.
- Smart industry is an interconnected network involving customers, OEM's, suppliers, partners, etc. since the OEM's need to collect, analyze and understand the data from all the stakeholders, this would become a threat to the OEM.
- Robust networking solutions and technologies are needed to share huge amounts of data from multiple devices in real-time. As the number of devices connected to industrial networks rises, and the volume of data they produce increases, the capacity of these networks can become slow.
- Network nodes and the topology of their networks needs to be done defined carefully, so that wiring can be managed efficiently paying attention to cost and complexity.
- Defining standards is a huge challenge to overcome, as the number of stakeholders involved is numerous, deriving a standardized system that works for all parties allowing both horizontal and vertical integration is a major challenge.
- Hackers could tap into smart factory networks and steal valuable information about customers, product designs or production processes. This can cause commercial damage by unfairly helping competitors, while also invoking claims from customers whose sensitive information is being stolen.
- Remote maintenance by equipment suppliers or subcontractors creates another potential risk, as it requires a connection to their network and computers.

- Upgrading existing production machines which lack digital identification and authentication functionality.
- Recruiting and developing new skills in human resources.

OODA in Industry 4.0

OODA is a strategy and decision-making model that focuses on observing data and detecting events from Industry 4.0, it determines the sense, filters the noise, reorients new information, and takes action all in a continuous iterative loop. The OODA stands for observe, orient, decide, and act, which is named after retired USAF Colonel James Boyd.

THE OODA MODEL

OBSERVATION: Observation relates to the data generated by humans, machines, and data from internet of things of a digital factory.

ORIENTATION: Orientation focuses on developing algorithms that will process the data and will predict the likelihood of results.

DECISION: By observing and orienting, a prioritized set of options are presented. Now by using predictive analytics and models the user can pick the course of actions, which will be best action from a list of potential actions.

ACTIONS: In action phase, execution of a business process transaction or function happens to produce a business result. These are actions taken in the implementation and application of a decision in a particular context within the digital factory. Here the user acts upon a prioritized list of decisions, also known as the Next-Best-Action, in a particular context or situation.

Quality management in Industry 4.0

Quality management in Industry 4.0 doesn't replace the traditional quality methods, but rather digitizes the traditional methods.

Smart quality management relies on:

- Management systems
- Technology used in data
- Technology used in software systems
- Technology used in infrastructure and its connectivity,

Management systems

Management systems in Industry 4.0 is built up on connected processes like risk management, compliance management, continual improvement, auditing, etc. to manage the end to end industry environment and people.

Quality control and assurance on Data

Data will proliferate in Industry 4.0, because machines, sensors, people, and many more things would generate data. This data would be of different types structured, unstructured, and semi-structured which can be of great speed and from different sources.

Hence quality control measures are necessary to verify and validate:

- Data coming from the right sources
- Data available in the right structure
- Data processed correctly
- Data being sent to the right person
- Data being stored at the right place

- Data share over the network between the authorized clients
- Data protection from risk and danger
- Events, trends and patterns of data
- Data predictions
- Data prescriptions

Apart from quality control, quality assurance is also needed with respect to process, policies, and operating procedures to handle the data.

Quality control and assurance on Software systems

There will be different types of software that will be used in Industry 4.0 which can be mentioned as software systems for manufacturing, collaborative systems, big data analytics, cloud systems, robotics, blockchain, social media, applications and many more.

Quality controls in these software systems will ensure

- To detect the changes in the software
- The usability of software with same functionalities in different environments and operate despite abnormalities in the network.
- To log the errors and failures in the software and report them to the OEM's for resolving.

Quality assurance is needed on software systems with respect to process, policies, standard operating procedures to manage them with respect to reliability, maintainability, confidentiality, stability, scalability, performance, availability, usability, and interoperability.

Quality control on Infrastructure and its connectivity

Infrastructure components in Industry 4.0 is humungous it ranges from a small sensor to a huge autonomous robot.

Quality controls in infrastructure and its connectivity will ensure:

- To select the most appropriate design, to assure that the user expectations are met.
- To identify and pinpoint defective or faulty parts before they fail.

Quality assurance is needed in infrastructure and its connectivity to ensure the infrastructure devices are safe, reliable, maintainable, confidential, stable, portable, robust, scalable, flexible, extensible, usable, and interoperable.

Quality programs in Industry 4.0

Quality programs that can be implemented in Industry 4.0 are: Total quality management, Six Sigma, 5S (sort, stabilize, shine, standardize, & sustain), and ISO 9000.

Total quality management (TQM) is a management approach for providing quality that emphases on accomplishing customer requirements and providing customer delight and satisfaction (for external and internal customers), motivating the involvement of the complete workforce, and continuous service improvement.

Customer satisfaction is the central focus of TQM, and products manufactured in Industry 4.0 should be in sync with the customer requirements, and it must be free from defects.

Six Sigma is a strategy and management improvement practice to expand growth and improve productivity of industrial organization. Six sigma launches quantifiable milestones for quality based on the number of standard deviations away from the mean in the normal distribution. This philosophy focuses on defect reduction and cost reduction, six sigma results in at most 3.4 defects per million opportunities. Here a defect is a measurable outcome that is not within acceptable limits and which lacks customer satisfaction.

Six-sigma practice involves in defining the problem, measuring, analyzing, recommending improvements, and developing a control plan to implement improvements.

Inspection: Inspection is a development and improvement procedure which comprises the usage of measurement and evaluating methods to control and regulate whether a service or an end-product is as per the design specifications.

Inspection is performed before, throughout, and after manufacturing. The incoming materials and starting parts are inspected upon receipt from suppliers; work units are inspected at various stages during their production; and the final product should be inspected prior to shipment to the customer.

Lean methodology:

Lean methodology is another manufacturing methodology that seeks to lessen and diminish the capital investments required to yield an industrial product or service.

In laymen words, the value-added time through a process should dramatically outweigh the non-value-added time. Lean methodology is all about eradicating unnecessary aspects from the industrial manufacturing process.

5S Methodology

5S is a method for creating and maintaining industrial infrastructure in an organized, disciplined, clean, and high-performance culture. 5S is the basis for continual product improvement, zero defects, cost reduction, and a safe work area. 5S is a systematic way to improve the industrial workplace, processes, operations and products through management and employee involvement.

5S stands for sort, stabilize, shine, standardize, sustain whose objectives can be defined as:

- Sort – Clearly distinguish needed items from unneeded items and eliminate the latter.
- Stabilize – Keep needed items in the correct place to allow for easy and immediate retrieval.
- Shine – Keep the work area clean.
- Standardize – Develop standardized work processes to support the first three steps.
- Sustain – Put processes in place to ensure that that the first four steps are rigorously followed.

Kaizen

Etymology of the word Kaizen comes from Japanese language which means "Kai – To modify + Zen – Make better".

This industrial management improvement methodology is to reduce lead times, machine failures and improve delivery performance, quality and delivery. It is particularly suitable for small scale improvements, risks are lower and can be obtained using a team-based approach. Kaizen assembles small cross functional teams aiming to improve a process or problem in a specific area.

Poka-yoke

Poka-yoke is a structured methodology for mistake-proofing operations. It is method that uses sensor or other devices for catching errors that may pass by operators or assemblers.

Poka-yoke effects two key elements of ZDQ (Zero Defect Quality):

- Identifying the defect immediately
- Quick Feedback for Corrective Action

The goal of poka-yoke is both prevention and detection. There are three approaches to Poka-yoke: Warning, Auto-correction, and Shutdown.

Efficient operations in Industry 4.0

Effective and efficient operations can be executed in Industry 4.0 by eliminating the wastes as mentioned below:

Over-deployment of different sensors in machines or deployment of cobots in industrial environment which creates confusions and chaos for the operational activities. Examples are: Deployment of unnecessary sensors (temperature sensors, proximity sensors, optical sensors, etc.) deviating from the functionality of a machine.

Collection of unnecessary data unnecessarily consumes network bandwidth, cloud/ hard-disk drive storage resulting in excessive and unnecessary costs for the industry. With the deployment of sensors on machines, machines would be generating data and will send it on network and gets stored on cloud or HDD.

Overproduction is producing more number of goods than the customer requirements. Examples are:

- Over-ordering materials
- Over producing products or goods

Correction is about fixing mistakes or doing the same job more than once, which is very common in manufacturing. Examples are:

- Making bad product
- Paying wrong vendor

Inventory is the liability of materials that are bought, invested in and not immediately used. Waste of Inventory is identical to overproduction except that it refers to the waste of acquiring raw material before the exact moment that it is needed.

Examples are:

- Overstocking raw materials

Unnecessary movement of people and equipment is inefficiency in industrial environment which includes looking for things like documents or parts/ equipments/ machines as well as movement that is straining.

Examples are:

- Movement of machines here and there without planning

Waiting can be due to slow or broken machines, material not arriving on time, etc. this inefficiency can cost huge monetary and reputation loss.

Examples are:

- Waiting for raw materials

Fishbone diagram for Industrial Operations

Fishbone diagram or Cause and effect analysis diagram is a problem analysis tool that can be used to analyze the issues in any industrial operations. This diagram represents a fish bone depicting the causes and effects for a specific problem statement. This method breaks down the problem into sequential layers of detail that potentially contribute to a particular effect. It also helps us to work on each cause prior to finding the root cause.

This diagram can be constructed by simple approach by defining the problem, brainstorming, and identifying the causes.

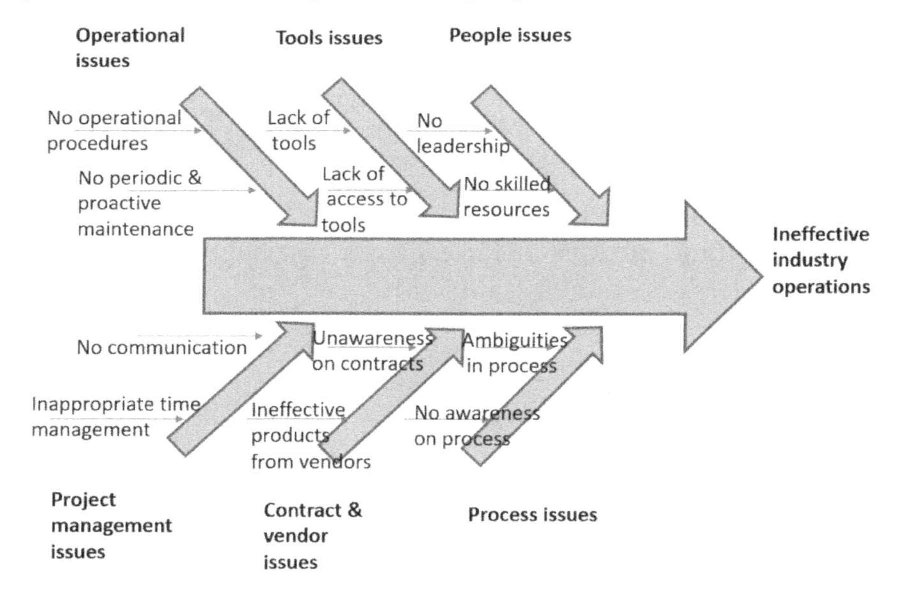

Fishbone diagram for ineffective industrial management operations can be represented by the below mentioned causes:

- **People related issues**

  People related issues can be segregated into issues like *high attrition, no motivation and encouragement for staff, no skilled and knowledgeable resources, lack of leadership, etc.*

- **Process related issues**

  Process related issues can be segregated into issues like *ineffective processes (which doesn't cover the complete scope of work), defects in processes, ambiguities in processes, no awareness on processes, etc.*

- **Tools related issues**

  Tools related issues can be segregated issues like *lack of Industry 4.0 machines, tools, software, KB tools, Lack of product/ tool knowledge and its functionalities, etc.*

- **Contract & vendor related issues**

  Contract & vendor related issues can be classified into issues like *Unawareness on contracts, inappropriate contract definitions, etc.*

- **Operational issues**

  Operational issues can be classified into issues like *no awareness on usage of technology or machines, no monitoring, repairs and failures, etc.*

- **Project management issues**

  Project management issues can be classified into issues like *no periodic training on process and technology, no communication plan in place or no communication happening with other process owners and operational stakeholders, inappropriate time management, inappropriate delegation, etc.*

Energy in Industry 4.0

Energy plays a very important role in industries across the world, and digitalization gives many advantages and opportunities for stable, available, continuous, higher efficient, and most importantly cost-effective power supply.

Without energy supply no industries work, and in the last 40 years energy consumption has raised to 40% and energy costs by 20%. Energy consumption in industries shows very clearly that power supply is consumed from different sources like fossil fuels, oil, natural gas, solar, nuclear power, etc. which also produces CO2 emissions.

But if the industries are supplied with hydroelectricity or renewable energy, it reduces CO2 emissions and also reduces the need to buy electricity from other sources. Surplus energy produced from hydroelectricity can also be sold to others which will also generate some income.

Hence usage of renewable energy is the solution to protect climate, to have better quality of energy supply, and reduce the costs.

To support renewable energy, there are different technologies like intelligent grids, distributed energy systems, integrated power supply to enable the stable, available and continuous power supply and will also ensure the industry is a green industry.

Intelligent grids

Intelligent grids can flexibly manage bidirectional power flow, intermittency, ensuring the entire system and its operations are safe and secure.

These grids can take new capacities, scalable with new technologies and can also integrate with renewable energy into existing grids to make the industrial infrastructure flexible for better operations.

Distributed energy systems

DES technology provides full independence from the grid, local resilience, and can manage demand to reduce peak loads when infrastructure is nearing capacity avoiding grid reinforcement investments.

With the help of DES there would be less impact on air pollution and will follow the standards defined by green industry.

Integrated power supply

Integrated power supply technology provides power distribution across all voltage levels with highest power quality ensuring energy efficiency.

Energy management in Industry 4.0

Availability and reliability of energy is crucial, and any kind of outages would lead to issues in business continuity, costs, and safety in industries. Hence here is my idea to implement effective energy management in Industry 4.0:

1. Perform as-is operations to understand energy flows and consumption.
2. Identify which assets/ machines are consuming excessive power.
3. Check the alternates or technology to reduce power consumption
4. Identify which assets/ machines are unnecessary or obsolete.
5. Develop awareness programs to save electricity.

6. Implement technology to monitor the power consumption and environmental impact so that employees can view it regularly.

CSF's for Industry 4.0

CSFs (Critical Success Factors) are those foundational pillars that are derived from the vision and mission statement of the organization/department; the factors mentioned below will help management (operational managers, tactical managers, high level managers & process managers) in defining a roadmap for the transformation to Industry 4.0.

Change for improvement of services

Companies that are resistant to change in implementing smart technology will lose market to those companies that change their business models to take advantage of the cloud, bigdata technologies, platforms and their new opportunities.

Manufacturing equipment like hardware, instruments, and heavy equipment are added with sensors and connectivity to their products, enabling predictive maintenance, security, and frequent upgrades. Sensor-based machines use data analytics to continuously gather data and make business decisions to improve efficiency in Industry 4.0 operations.

Strategy implementation across all your value chain

Management should implement their strategy (automated, smart machinery) across all the value chain which typically consists of inbound distribution or logistics, manufacturing operations, outbound distribution or logistics, marketing and selling, after-sales service, purchasing or procurement, research and development, human resource development, and corporate infrastructure.

Standards, policies and plans should be implemented across all your value chain, based on your new strategy suitable for smart machines and infrastructure. New strategy should focus on creating a new technological base on which a wide range of vendors and customers can interact seamlessly with the same collection of hardware, software, services, and one another.

Build a customer centric industry

Smart industry working model relies on the web of connections between manufacturers, retailers, and consumers. Today, as many customers and consumers are able to communicate directly to manufacturers of the products and services, industries should set up their factories in such a way that the factory can receive the customers' preferences and rapidly convert them into designs.

Develop the necessary technological knowledge

Industries become smart with the latest and innovative technologies like augmented reality, working with cobots, analyzing data, etc. hence recruiting people or training people with software expertise, and raising the technical skills of all employees, partners, etc. is a very essential activity.

Implement in a phased approach

Implementing smart machines, smart equipment, big data and analytics in industry should be done in a phased approach (department by department, geographical location by location, etc.) as steady stream can be easier to implement, test, understand the pros and cons, its profitability and see customers response before rolling them out worldwide.

Continuously analyze your data

Continuously analyzing industrial data from all your operations and customer insights is critical to use the analytic results to recognize important patterns, and to gain insights that help you make the right choices and keep improving the industrial operations.

Skills needed for Industry 4.0

With the advent of Industry 4.0, employees/ workforce should also upgrade their skills with respect to the new and changing demands.

In the late 1930-50's it was the time where mechanical engineers were in great demand and that was the time when innovation just started to creep in electrical & electronical sciences.

In 1950-1960's, the trend was towards electrical and electronical engineering. After few years it moved to computer sciences and now with the fourth industrial revolution, people are anxious to know what is the new demanding skills that will make their careers and jobs safe.

Well, the good thing about Industry 4.0 is it requires all kinds of engineering skills with respect to mechanical, civil, electrical, electronical, computer programming, artificial intelligence & environmental sciences.

It's a misconception that in Industry 4.0 there will be no workforce/ very less workforce. Without human workforce no industry existed in the past and will never exist, workforce will have to upgrade themselves with new technologies to work in a smart way.

The skills that will be in demand for Industry 4.0 are:

CAM/ CAD/ CAID/ CAE/ CAS Expertise:

Computer aided designing and manufacturing will be vital in Industry 4.0, as it will be used to create, modify, analyze or optimize engineering design.

Mechanical engineers with expertise in CAM/ CAD/ CAID/ CAE/ CAS will be ideal for mechanical engineering jobs.

Design and automation engineering expertise:

Design and automation engineering expertise is required to

- design, manufacture, produce, test, perform operations and maintenance.
- develop RF architecture and system design, UHF RFID readers, producing a RFID communication system on hardware, configuring RFID hardware including readers, antennas, and peripherals.
- program automated machinery using Programmable Logic Controls (PLC) and configure Human Machine Interface (HMI).

Electrical and electronical engineers with expertise in sensor manufacturing, RFID, RTLS, HMI, etc. will be ideal for design and automation engineering jobs.

IoT engineering expertise

IoT engineering expertise is required to:

- collaborate to develop system architectures, data transport schemas, and deployment strategies for existing and new infrastructure.
- achieve seamless integration of wireless sensors, edge nodes, gateway servers, and routers within hybrid networks.
- research and select equipment suppliers, system integrators, and service providers and work closely with them on resolution of various design and production issues.
- work with third party consultants to meet the necessary regulatory requirements, as required by the product specifications.
- setup and test instances of IoT solutions comprised of hardware and software, gateway servers and sensors.

Electrical and electronical engineers with expertise in (wireless technologies, network architecture, sensors, gateways, etc.) will be ideal for IoT engineering jobs.

Big data and analytics expertise:

Bigdata and analytics expertise is required to

- architect, design, adopt and apply big data strategies and architectures.
- develop an overall IIoT and Mobile ecosystem engagement strategy.

Computer science/ IT engineers with expertise in (Big Data systems, ETL, data processing and analytics, Hadoop, spark, SQL, Hive, etc.) will be ideal for big data jobs.

Cloud computing expertise

Cloud computing expertise is required to

- operationally support activities like platform monitoring, creation and maintenance of automation scripts.
- design cloud models and service infrastructure based on the requirements.
- develop applications within the cloud platform that will connect smart sensors, edge devices, gateways and RF infrastructure for smart living.

Computer science/ IT engineers with expertise in Java, vSphere / ESXi / AWS / OpenStack / Azure / GCP, MQTT (protocol) will be ideal for cloud computing jobs.

Robotics expertise

Robotics engineering expertise is required to

- design multi-faceted robots and cobots based on unmanned ground and aerial vehicles.
- develop high-level behaviors and logic for coordination of multiple assets.
- test, troubleshoot, and document performance of algorithms, communications networks, and robotics hardware.

Computer science/ IT engineers with expertise in (C++, python, object-oriented programming and robotics design, integration, sensors, control, etc.) will be ideal for robotics engineering jobs.

Cybersecurity expertise

Cybersecurity expertise is required to

- define, implement, and model security practices for enterprise environments using an intelligent and threat-driven defense model.
- develop documented security architectures detailing secure, defendable configurations of systems, applications, and processes.
- support businesses in designing, implementing and deploying secure systems.
- validate designs for security throughout their lifecycle.
- identify and assess risk and make sound risk decisions and recommendations.

Computer science/ IT engineers with expertise in (Cryptography, security, privacy, software development (C/C++ and Java), and/or embedded systems, etc.) will be ideal for machine learning engineering jobs.

Machine learning expertise

Machine learning expertise is required to

- identify new ways to co-design architecture and hardware-centric algorithms and the right tradeoff to reduce complexity and power consumption as needed.
- design architecture of hardware block and software interfaces.
- evaluate architectures in terms of performance, power consumption, area, complexity and flexibility.
- develop a simulation reference model for software and applications.
- develop accelerator infrastructure for integration into sensors and the computer system.

Computer science/ IT engineers with expertise in (C, C++, Java, PHP, machine learning, mobile platforms, prediction and machine learning, algorithms, etc.) will be ideal for machine learning engineering jobs.

Countries approach on Industry 4.0

Today many countries across the world have their focus and attention on Industry 4.0, and developed countries like USA, Germany, UK, Japan and China are very much ahead in the race of fourth industrial revolution.

So here is a glance on the countries, their perspective and their approaches for Industry 4.0.

USA

USA a developed super powerful country with GDP 19.390 trillion USD, and manufacturing contributes 15% to its GDP. USA has developed various initiatives to provide a direction for Industry 4.0 like:

- Development of NNMI (National Network for Manufacturing Innovation) which is focusing on various aspects like additive manufacturing, 3D printing, modern metal manufacturing, low cost energy efficient manufacturing of fiber-reinforced polymer composites, etc.
- Materials genome initiative to discover, manufacture, and deploy advanced materials twice as fast.
- National robotics initiative to develop robots that can work in industries.

Germany

Germany another developed country with GDP 3.7 trillion USD, and manufacturing contributes 30% to its GDP.

- Germany was the first country who talked and initiated about the fourth industrial revolution in 2011.

- Today there are many factories which can be called as smart factories like BMW, Robert Bosch GmbH, Daimler AG, SAP SE, Siemens AG, and many more.
- There are more than 50 universities which are offering bachelors and masters programs specializing in Industry 4.0 technologies.

India

India a developing country with GDP 3 trillion USD, and manufacturing contributes 29% to its GDP. Currently awareness on Industry 4.0, in India is in the moderate stage; hence government has taken great initiatives like:

- Fraunhofer-Gesellschaft will support the Indian government as a Technology Resource Partner and will create a roadmap for technological development of the Indian manufacturing industry.
- 'Smart Cities Mission', building 100 smart cities across India as the forerunners of the Industry 4.0 environment.
- Indian Institute of Science (IISc) building India's first smart factory in Bengaluru.
- Bosch, a German auto component manufacturer which will begin implementation of smart manufacturing at its 15 centers in India by 2018.
- General Electric to develop multi-modal factory in India where digitally interlinked supply chains, distribution networks, and servicing units form part of this intelligent ecosystem.

- AlfaTKG a Japanese technology firm to work with the Indian Institute of Technology, Chennai to undertake research for developing smart manufacturing technologies for India.

China

China a developed country with GDP which is close to 14.2 trillion USD and manufacturing contributes 35.37% to its GDP. Currently implementation of IoT and digitization in industries is at a very good pace.

PRC (Public Republic of China) government has taken extraordinary initiatives and is supporting Industry 4.0 initiative with:

- Made in China 2025 policy, to significantly improve the overall manufacturing quality, enhance creativity and productivity, and integrate industrialization and information.
- A mobile phone module manufacturer (Changying Precision Technology Company) which is country's first unmanned factory.
- Ningbo Techmation (R&D center), to make robots and industrial automation equipment.

Brazil

Brazil a developing country with GDP which is close to 3.2 billion USD and manufacturing contributes 18% to its GDP. Currently awareness on Industry 4.0, in Brazil is in the early stage hence government has taken great initiatives like:

- Trainings on latest Industry 4.0 technologies to university students.
- Virtual trainings to workforce in industries.
- Development of new policies for smart manufacturing.

- Collaboration with Ericsson to drive innovation and implement smart devices in the country.

Russia

Russia a developed country with GDP which is more than 1.3 trillion USD, and manufacturing contributes 32% to its GDP.

Currently awareness on Industry 4.0, in Russia is in the early stage hence government is supporting great initiatives like:

- 'Development of manufacturing industry in the horizon 2020' this program focuses implementing Industry 4.0 components in the next 5 years. Goals have been set to produce all materials required for robotics and nano-electronics locally, which will be a significant shift from the current import of 90% of these materials.
- Subject on Robotics included in schools and universities, close to 30 universities are offering robotics engineering to university graduates.
- ABB Electrical Group is developing a robotics center in Technopolis Moscow. This center will develop technological solutions for robotized systems of arc and spot welding and high-precision material processing adapted for the Russian market.
- Intel opening-up an Ignition Lab to provide solutions to Russian companies using intel processors for transport and energy.
- Mitsubishi Electric operations in Russia which was started in October 2014 to strengthen Mitsubishi's factory automation business.
- SAP has opened Russian Center for IoT in 2016 to provide customers and partners an opportunity to learn about the company's IoT technologies.

South Africa

South Africa a developing country with GDP which is more than 340 billion USD, and manufacturing contributes 30% to its GDP.

Currently awareness on Industry 4.0, in South Africa is in the early stage hence government has started and is supporting great initiatives like:

- Research programs related to additive manufacturing in collaboration with The Department of Science and Technology (DST).
- Promoting development of advanced manufacturing technologies in the areas of additive manufacturing, automation, advanced electronics, photonics and aero-structures in collaboration with TIA (Technology Innovation Agency).
- Establishment of SEZ's to attract advanced foreign production and technology methods in order to gain experience in global manufacturing.

Frequently asked questions

Q. Is Industry 4.0 just a buzzword or a reality?

A. Industry 4.0 is definitely a reality which has started in European countries and will progress rapidly throughout the world in the next few years. For now, may be few companies have defined their strategy and are transforming themselves into smart factories. But very soon many industries would become desperate to make their factories as smart factories, since the benefits that they would get from a smart factory are numerous.

Q. What is a smart factory?

A. Factory that is connected, proactive, agile, adaptable, and has good resource efficiency with the help of latest technologies like robotics, AI, sensors, cloud, etc.

Q. What does connectivity mean in smart factory?

A. Connectivity in smart factory means having real-time data-enabling collaboration with all the stakeholders and being able to continuously pull data with sensor and location-based services.

Q. Why do we need autonomous robots and cobots in Industry 4, why can't we have human workforce?

A. Autonomous robots and cobots will be used like tools, and they would never replace human staff. These tools will be used for the routine and repetitive work tasks, which will make the human workforce lull and dull doing the same routine job and will kill the passion in their jobs. These robots and cobots are like plug n' play solutions, easy to deploy, easy to program and have less maintenance with the help of prescriptive and predictive maintenance.

Q. Security is a big concern in Industry 4.0, how can we mitigate it?
A. Having good information security policies and processes, identity and access management tools, and biometric tools in place would be a good start. But regularly reviewing and auditing them is a must. The real threat is from insiders, so we need to ensure that we keep the data, systems, and machines secure with the right tools.

Q. How should I transform my company to smart company?
A. Firstly, identify the routine tasks in your organization use automation to make them faster and efficient.
Get the confidence of your customers and showcase the benefits.
Use MTconnect protocol, so that machines can send real time details about their machines.
Do not go in a big bang approach, try doing it in a phased approach department by department or location by location.

Q. Why should I make my factory as a smart factory?
A. Smart factory enables your operations to increase quality, reduce processing & manufacturing time and also reduces costs.

Implements digitization which consolidates all information in the factory processes, which can be analyzed and transformed to provide meaningful insights about the Industry processes.

Q. How much time does it take to 3D print any object?
A. It all depends on the technology that you are using. But generally, the time taken to print depends upon the size of the object and settings set for an object. A small object with less dimensions and infills can be printed in less time than the objects with higher dimensions.

Q. Augmented reality is very expensive, I am afraid my factory can't afford it.
A. Augmented reality is a worthwhile investment for your factory operations. Also remember that there is a wide range available. If you are looking for the best quality, then you will have to invest good money which will definitely give you the ROI.

Q. How do I know, which kind of sensors would my machines need?

A. Understand the machine conditions and constraints like temperature, vibrations, pressure, etc. Accordingly, test the sensing technology that meets your machine and infrastructure requirements as a POC (Proof of concept) and then purchase the sensors.

Q. How different is SCADA from IOT?
A. IoT is methodology which makes devices smart and SCADA is one such technology to make your device smart. Information generated from SCADA systems acts as one of the data sources for IoT. SCADA's focus is on monitoring and control. IoT's focus is firmly on analyzing machine data to improve your productivity and impact your top line.

Q. Can small factories aspire to become smart factories?
A. Yes, every organization wants to provide better service to customers, reduce costs, build better brand and make better profits.

Here are some of the reasons why small factories can afford to become smart factories:

- Firstly, sensors using Radio Frequency Identification (RFID) technology have become cheap.
- Secondly, predictive analytics, cognitive computing and artificial intelligence can make effective decisions and predictions based on sensed data.

- Thirdly, the human-machine interface allows intuitive interaction between physical and digital worlds.
- Fourthly, the ability to simulate a product is possible through 3D printing with new flexible system of production which provides a new flexibility to see a virtual product.
- With all the above mentioned, customer gets better confidence on the company and will have the transparency on the product development lifecycle.

Q. What is the difference between IOT and Industry 4.0?
A. The Internet of Things (IoT) is branch in IT where all things will be connected through internet like people, processes, data and things. (IoT is a component of Industry 4.0.)
Industry 4.0 is next industrial revolution and new approach for running industries using smart machines, smart infrastructure which enables connectivity of machines and equipment on the factory floor.

Q. What is the difference between Industry 4.0 and 3.0?

A. Industry 4.0 is focused on hyper-connectivity, where every machine and its parts are connected using latest technologies like sensors, gateways, networks, cloud, etc. along the value chain where customers, suppliers, logistics, etc. are involved.

Industry 3.0 was focused on automation using computers, robotics and software.

Q. Which technology will play significant role in Industry 4.0?

A. There are many technologies that are involved in Industry 4.0, like AI, blockchain, 3D printing, Cloud, etc. Every technology has its own prominence, so no technology can outweigh another.

Q. Which industries are quick at implementing smart factories?

A. Automotive and Aerospace are the two industries that have transformed their factories to smart factories in the last few years.

Q. Will Industry 4.0 reduce the human workforce?

A. Today taxi service has got digitized by some companies like uber, grab, and ola now that doesn't mean these companies have removed human workforce. They are using digital technology in the taxi service, but still the drivers are humans and it has provided great employment opportunities.

In the same way Industry 4.0, will have latest infrastructure which can communicate, process, work with digital information but still needs human intelligence to make those machines work.

Q. What is the difference between Bigdata and Artificial Intelligence?

A. Big data is a practice of storing huge volumes of data and AI (machine learning, deep learning, etc.) is a technology and methodology that works on this huge volume of data to make effective decisions.

Q. How to bring awareness in Industry 4.0?

A. Awareness on the next industrial revolution is the most important thing for gearing up towards Industry 4.0, here are some points to bring awareness for industrialists:

- Strengthen employee training: The workplace has to provide good number of trainings as per the employee's qualifications and skills.
- Create the conditions to facilitate the required training: Providing trainings to schools, universities and other

education institutions on the skills needed for Industry 4.0 at all levels – regional and national level. Schools and higher education institutions have to teach media and digitalization skills in order to ensure that school children and students are properly equipped for Industry 4.0.

- Use digital media to teach Industry 4.0 skills: Digital learning formats and methods like seminars, workshops, and tutorials will be important to develop need-based, self-guided, informal and formal solutions.

Appendix

Acronyms

ADF – Adaptive Digital Factory

AMP – Advanced Manufacturing Partnership

AMR – Autonomous Mobile Robots

ANN – Artificial Neural Networks

APT – Advanced Persistent Threat

AR – Augmented Reality

AV – Autonomous Vehicle

BIA – Business Impact Analysis

BIM – Building Information Modeling

BOM – Bill of Materials

CAN – Control Area Network

COTS – Commercial of the shelf

CPS – Cyber Physical Systems

CSI – Continual Service Improvement

DFSS – Design for Six Sigma

DMAIC – Define, Measure, Analyze, Improve and Control

DMADV – Define, Measure, Analyze, Design and Validate

DSN – Digital Supply Network

GIS – Geographic Information Systems

HMI – Human Machine Interfaces

IAR – Industrial Augmented Reality

ICS – Industrial Control Systems

IIC – Industrial Internet Consortium

IIoT – Industrial Internet of Things

IoA – Internet of Automation

IoR – Internet of Robotics

IoT – Internet of Things

IoS – Internet of Services

IPS – Intelligent Positioning System

IWSN – Industrial Wireless Sensor Networks

IVCI – Industrial Value Chain Initiative

KPI – Key Performance Indicators

M2M – Machine to Machine

MOM – Manufacturing Operations Management

MTC – Manufacturing Technology Centre

OEE – Operational Equipment Effectiveness

OEM – Original Equipment Manufacturer

OODA – Observe, Orient, Decide & Act

OT – Operational Technology

PLC – Programmable Logic Controllers

PLM – Product Lifecycle Management

PMQ – Predictive Maintenance Quality

PSM – Process Safety Management

RTD – Resistor temperature detectors

RACI – Roles, Accountability, Consulted, and Informed

RCA – Root Cause Analysis

RFID – Radio Frequency Identification

RTLS – Real Time Location System

ND – Natural Disaster

SCADA – Supervisory Control & Data Acquisition

SLA – Service Level Agreement

SOP – Standard Operating Procedures

SPOC – Single Point of Contact

KB – Knowledge Base

VBF – Vital Business Function

QC – Quality Control

QoS – Quality of Service

Index

Made in the USA
Middletown, DE
30 January 2020

83975860R00116